Learning Algorithms
Theory and Applications

S. Lakshmivarahan

Learning Algorithms
Theory and Applications

Springer-Verlag
New York   Heidelberg   Berlin

Prof. S. Lakshmivarahan
University of Oklahoma
School of Electrical Engineering
and Computer Science
Norman, Oklahoma 73019

Library of Congress Cataloging in Publication Data

Lakshmivarahan, S.
   Learning algorithms theory and applications.

   Bibliography: p.
   Includes index.
   1. Artificial intelligence.   2. Algorithms.
3. Learning.   I. Title.
Q335.L34   001.53′5   81-16683
                                AACR2

© 1981 by Springer-Verlag New York Inc.
Softcover reprint of the hardcover 1st edition 1981

9 8 7 6 5 4 3 2 1

ISBN-13: 978-0-387-90640-9   e-ISBN-13: 978-1-4612-5975-6
DOI: 10.1007/978-1-4612-5975-6

DEDICATED TO

MY

FATHER and MOTHER

Who taught me many useful algorithms.

PREFACE

Learning constitutes one of the most important phase of the whole
psychological processes and it is essential in many ways for the occurrence
of necessary changes in the behavior of adjusting organisms. In a broad sense
influence of prior behavior and its consequence upon subsequent behavior is
usually accepted as a definition of learning. Till recently learning was regarded
as the prerogative of living beings. But in the past few decades there have
been attempts to construct learning machines or systems with considerable success.

This book deals with a powerful class of learning algorithms that have been
developed over the past two decades in the context of learning systems modelled
by finite state probabilistic automaton. These algorithms are very simple iterative
schemes. Mathematically these algorithms define two distinct classes of Markov
processes with unit simplex (of suitable dimension) as its state space. The
basic problem of learning is viewed as one of finding conditions on the algorithm
such that the associated Markov process has prespecified asymptotic behavior.
As a prerequisite a first course in analysis and stochastic processes would be
an adequate preparation to pursue the development in various chapters.

Automaton models and algorithms for learning, which is the central theme of
the present book, have been widely accepted as one of the fundamental approaches
to machine intelligence. In spite of this wide spread acceptability most of the
published material on this topic to date are confined to articles in research
journals or isolated chapters in anthologies and as such there is no one single
text book or monograph on this topic. This book is intended to fill this gap.

The book consists of two parts: Part I developes the theory of design of learning algorithms. In particular it deals with the convergence behavior of various classes of learning algorithms. Part II describes the application of these algorithms to decentralized decision making problems with incomplete information. More specifically, we deal with the applications to two person zero sum sequential games with incomplete information, two person team problem with incomplete information and Markov decision problem with incomplete information.

Our aim is to provide a unified framework and a comprehensive treatment of most of the results pertaining to both the theory and applications known to date. Interestingly enough, much of the results presented in Part II came into existence only during the past three to four years.

Basically, the book arose out of the notes written by the author for a series of lectures delivered at the Institute of Applied Mathematics, University of Bonn, W. Germany during January 1980 under the auspecies of their Sonderfor-schungsbereiches. A major part of an earlier version was also used for the graduate course CS 5070, Seminar in Problem Solving at the School of Electrical Engineering and Computer Science, University of Oklahoma during Spring Semester 1980.

The author wishes to record his sincere appreciation and thanks to Professors R. Leis, W. Vogel, V. Herkenrath and D. Kalin of the University of Bonn for arranging this visit and their hospitality. Special thanks are due to Professors Herkenrath and Kalin with whom the author has spent countless hours discussing various aspects of both the theory and applications of learning algorithms. Dr. Herkenrath read through the entire manuscript and has suggested various improvements. It was Professor Radu Theodorescu of Laval University who introduced the author to the University of Bonn. Our thanks are due to him and his collaborator

Professor M. Iosifescu for their constant encouragement. Professor M.A.L.
Thathachar initiated the author into the field of learning algorithms when
he was a graduate student at the Department of Electrical Engineering, Indian
Institute of Science, Bangalore, India. Ever since that time, he has been a
great source of inspiration and continued guidance. His influence will be felt
throughout the book and the author is deeply indebted to him. Much of the work
in Part II (Chapters 5 through 7) was carried out when the author was at Yale
University during 1976 - 1978 under the direction of Professor K.S. Narendra.
It was Professor Narendra who introduced the author into the exciting area of
learning approach to decentralized decision making with incomplete information.
To him we are very grateful for his continued advice and support. In addition,
quite a variety of people have helped the author in many different ways. Professor
H.J. Kushner helped to clarify many intricacies of different classes of learning
algorithms. In fact, his well known counter example marked a definite turning
point (for the better) in our research work. Professors K.S. Fu, M.D. Srinath
and H. Aso have read through the various portions of the manuscript and have
suggested various modifications. Professors M.E. Council and S.K. Kahng (past
and present directors) of the School of Electrical Engineering and Computer Science,
University of Oklahoma have extended full support for the completion of this work.
The Mathematics editorial staff of Springer Verlag, Inc. New York handled this
project with great care and patience. Our sincere appreciation and thanks are
due to all these above individuals. Thanks are also due to Betty Sudduth,
Vici Chance and Ronda Sharp for typing various portions of the manuscript.

Finally, to all members of my family, especially my wife Shantha and our
children, Subha and Bharathram for their understanding and help (too numerous to

list) words are not enough to adequately thank them.

Norman, Oklahoma

July 1981   S. Lakshmivarahan

TABLE OF CONTENTS

PART 1, THEORY

PART 1

THEORY

CHAPTER 1

INTRODUCTION

1.1. Various Approaches to Learning:

Learning has been the interest of psychologists and mathematicians for decades and more recently, of engineers and computer-scientists. The interest
of a psychologist or a mathematician in learning is to explain or describe
the way in which animals learn to do a variety of skills by observing the
changes in their behavior. Such an approach may be termed as the descriptive
approach. A wide range of mathematical models have been developed for this
purpose. The work by Bush and Mosteller [B10], Luce [L14] Hilgard [H4],
Norman [N10], Iosifescu and Theodorescu [I2] to mention a few, belong to this
category. On the contrary, in systems theory and computer science the aim is
to develop a computer program or build a machine, perhaps in the context of
pattern recognition or artificial intelligence, which will learn to perform
certain prespecified tasks such as to play games or classify a class of x-ray
pictures and diagnose disease, etc. Fu [F4] [F5] [F6], Tsypkin [T4] [T5],
Sklansky [S9], Mendel [M6 [M7], Saridis [S4], Nilsson [N9], Csibi [C10] Slagle
[S10], Fukunaga [F8]. Such an approach is often called the prescriptive
approach to learning.

Learning is often associated with a goal or a performance measure. For lack of sufficient information, often the goal of learning may not be completely specified. In this context learning has a dual role: (1) compensate for insufficient information by appropriate data collection and processing and (2) in that process incrementally move towards the ultimate goal.

In Systems Theory and Computer Science learning has been implemented in various ways: using stochastic approximation methods [T4] [T5] [S3] [F4] [F6], heuristic programming techniques [N9] [S10], inductive inferential techniques [S13], statistical inference techniques [F5] [F8] and automaton models [T2] [N3] [N1] [C1] [L4] [M3] [M4]. In this book we will confine our attention to the study of algorithms that are developed in the context of automaton models. These algorithms are called learning algorithms.

1.2. A Learning Algorithm:

There are $M(\geq 2)$ coins. On tossing, a coin falls either head or tail. Let us call the outcome falling head a success and falling tail a failure. Let d_i be the probability of success and $c_i = 1 - d_i$ be the probability of failure for coin i, $i = 1, 2, \ldots, M$. We assume that all the d_i's are all distinct (i.e.) $d_i \neq d_j$ for $i \neq j$ and $0 < d_j < 1$ for all i,j and let $D \underline{\underline{\Delta}} (d_1, d_2, \ldots, d_M)^T$ where T denotes transpose. To make the learning problem non-trivial and interesting the following fundamental assumption is made:

<u>Assumption 1.1.</u>: The d_i's are all <u>unknown</u> and without loss of generality let $d_1 > d_2 > d_3 > \ldots > d_M$.

Let $p(k) = (p_1(k), p_2(k), \ldots p_M(k))^T$ where $p_s(k)$ denote the probability of choosing the sth coin at time k for tossing. The basic idea of the learning algorithm may be stated in words as follows: <u>Increase</u> the probability of choosing the ith coin at stage (k+1) if it was chosen at time k and the toss resulted in a success and <u>decrease</u> the probability of choosing the ith coin if it is chosen at time k and the toss resulted in a failure. The increase and the decrease are called <u>reward</u> and <u>penalty</u> respectively. The probabilities of choosing all the other coins are adjusted to keep the total probability equal to unity.

More specifically, let

$$S_M \triangleq \{p \mid p = (p_1, p_2, \ldots, p_M)^T, \ 0 \le p_i \le 1, \ \sum_{i=1}^{M} p_i = 1\} \qquad (1.1)$$

be the M-dimensional unit simplex, and

$$V_M \triangleq \{e_i, \ i = 1, 2, \ldots, M \mid e_i \text{ is the } i^{th} \text{ unit}^{(1)} \text{ vector}\} \qquad (1.2)$$
$$\text{of dimension M.}$$

Clearly, V_M corresponds to the vertices of S_M. Let $f_s^i[.]$ and $g_s^i[.]$ be continuous functions such that $f_s^i : S_M \to [0,1]$ and $g_s^i : S_M \to [0,1], i,s = 1,2,\ldots,M$.

The algorithm stated below captures the spirit of the verbal description given above.

---

(1) i^{th} unit vector is one whose i^{th} component equals 1 and the rest are all 0.

$$p_i(k+1) = p_i(k) + f_i^i \, [p(k)]$$

$$p_j(k+1) = p_j(k) - f_j^i \, [p(k)], \; j \neq i$$

if coin i was chosen and resulted in a success $\qquad$ (1.3)

$$p_i(k+1) = p_i(k) - g_i^i \, [p(k)]$$

$$p_j(k+1) = p_j(k) + g_j^i \, [p(k)], \; j \neq i$$

if coin i was chosen and resulted in a failure

where

$$f_i^i(p) = \sum_{j \neq \mathbf{L}} f_j^i(p)$$

$$g_i^i(p) = \sum_{j \neq \mathbf{L}} g_j^i(p)$$

$\forall \, i = 1, 2, \ldots, M$

$p \, \varepsilon \, S_M$ $\qquad$ (1.4)

The above algorithm is called <u>non-linear reward-penalty</u> N_{R-P} <u>algorithm.</u>[2] The p(k)'s as defined in (1.3) are all random vectors. As the vector D of success probabilities and the function $f_s^i(.)$ and $g_s^i(.)$, $s = 1, 2, \ldots, M$ are all independent of k, $\{p(k)\}$, $k \geq 0$ is a <u>discrete time Markov process</u> with stationary transition function over the State Space S_M. It is obvious that we need further conditions on $f_s^i(.)$ and $g_s^i(.)$ for p(k+1) $\varepsilon \, S_M$ for all $k > 0$ given that p(0) $\varepsilon \, S_M$. These additional conditions very much control the behavior of the Markov process $\{p(k)\}$. In fact, it is our aim to find these conditions such that $\{p(k)\}$ will exhibit certain prespecified behavior.

1.3. Performance Measures and Statement of Problem:

The required behavior of a learning algorithm (or the associated random process such as $\{p(k)\}$) is often specified in terms of various performance measures.

The average probability of success at stage k is given by

[2] The increase (decrease) in $p_i(k)$ when the toss of the coin i resulted in success (failure) is called reward (penalty).

$$\eta(k) = \sum_{i=1}^{M} P_i(k) \, d_i \tag{1.5}$$

Let

$$\eta_0 = \frac{1}{M} \sum_{i=1}^{M} d_i \tag{1.6}$$

Definition 1.1. The learning algorithm is said to be

    (a) <u>Expedient</u> if $\lim_{k \to \infty} \inf \, E[\eta(k)] > \eta_0$ (1.7)

    (b) <u>Optimal</u> if $\lim_{k \to \infty} E[\eta(k)] = d_1$ (1.8)

    (c) <u>ε-optimal</u> if $\lim_{k \to \infty} \sup \left| E[\eta(k)] - d_1 \right| < \varepsilon, \quad \varepsilon > 0$ (1.9)

        and

    (d) <u>Absolutely Expedient</u> if

$$E[\eta(k+1) \mid P(k)] \begin{cases} > \eta(k) & \text{if } p(k) \in S_M^o \\ = \eta(k) & \text{if } p(k) \in V_M \\ \geq \eta(k) & \text{otherwise} \end{cases} \tag{1.10}$$

With probability one, for all $d_s \in (0,1)$ $S = 1$ to M where $S_M^o = \{ P = (P_1, P_2, \ldots P_M)^T \mid 0 < P_i < 1 \text{ and } \sum_{i=i}^{M} P_i = 1 \}$ and $E[\cdot]$, $E[\cdot|\cdot]$ are the expectation and conditional expectation operators.

<u>Statement of Problem</u>: Our aim is to develop different classes of learning algorithms such that the Markov process $\{p(k)\}$ will satisfy one or more of the performance measures stated above.

Notice that the average probability of success is a linear functional defined on S_M. Stated in words, expediency requires the asymptotic expected value of average probability of success to be greater than that would   d be obtained by a pure random choice of the coins. Absolute expediency requires $E[\eta(k)]$ to be strictly monotonically increasing. Further, absolute expediency implies[3] expediency if $p(o) = (1/_M, 1/_M, \ldots, 1/_M)^T$. It will be shown in Chapter 3 that absolute expediency implies ε-optimality. It should be interesting to note that at the present time there is <u>no</u> algorithm that is known to be optimal, though a large class of ε-optimal algorithms will be presented in Chapters 2 and 3.

1.4 Classification of learning algorithms:

Let $(S_M, S_M, \mathcal{M})$ be a metric space where $\mathcal{M}$ is the Euclidean metric on the state space and S_M is the set of all open sets in S_M with respect to $\mathcal{M}$. Let $I = \{1,2,\ldots,M\}$ and $E = \{success, failure\}$. The learning algorithm (1.3) defines a mapping

$$T: S_M \times I \times E \to S_M$$

where

$$p(k+1) = T[p(k), i(k), e(k)] \tag{1.11}$$

where $i(k) \in I$ denotes the coin chosen and $e(k) \in E$ the event occurring at time k. E is called the <u>event</u> space, p(k) and e(k) are known as the state and event sequences respectively. Let T^* be an extension of T defined as follows:

$T^*: S_M \times (I \times E)^n \to S_M$ for all $n \geq 1$ where $(I \times E)^n$ refers to the n-fold cartesian product of $I \times E$ and

1) $T^* = T$ for $n = 1$

2) $T^* [p, \{(i_0,e_0), (i_1,e_1) \ldots (i_n,E_n)\}]$

$= T[\ldots T[T[p, i_0, e_0], i_1,e_1]\ldots, i_n, e_n]$ $\tag{1.12}$

---

where $i_\ell \in I$ and $e_\ell \in E$ for $\ell = 0,1,2,\ldots,n$, and $n \geq 2$.

Define

$$\mathcal{K}: S_M \times (I \times E) \to [0,1] \quad \text{as}$$

$$\text{Prob } [(i,e) \mid p] \triangleq \mathcal{K}[p,(i,e)]$$

and

$$\text{Prob } [\big(i(k+1),\, e(k+1)\big) \mid p(0) = p,\ (i(\ell),\, e(\ell)) : \ell = 0,1,\ldots,k] \\ = \mathcal{K}[p',(i(k+1),\, e(k+1))]$$

(1.13)

where

$$p' = T^*[p(0),\ \big\{(i(0),\, e(0)),\ (i(1),\, e(1)),\ \ldots\ (i(k),\, e(k))\big\}]$$

clearly $\mathcal{K}[p,(i,e)] = p_i d_i$ if $e = \{\text{success}\}$

$$= p_i c_i \text{ if } e = \{\text{failure}\}.$$

That is, $\mathcal{K}[p, \cdot]$ is the event probability distribution.

Definition 1.2: The quadruple $(S_M,\ I \times E,\ T, \mathcal{K})$ is called a (homogeneous) Random System with Complete Connection (RSCC). [I2] [N10]

Remark 1.1: Motivated by the coin tossing experiment we have assumed that d_i (or c_i)'s are all constants. However, in part II (chapters 5, 6 and 7) we will encounter applications in which d_i's in general depend on the state p as well. In this and in the rest of the chapters by a learning algorithm we implicitly refer to the associated RSCC $(S_M,\ I \times E,\ T,\ K)$.

Let

$$\gamma [T[\cdot,\ i,e]] = \sup_{\substack{p^1 \neq p^2 \\ p^1, p^2 \in S_M}} \frac{\mathcal{M}[T[p^1,i,e],\ T[p^2,i,e]]}{\mathcal{M}[p^1,p^2]}$$

(1.14)

Definition 1.3: A learning algorithm T is said to be <u>distance diminishing</u> if

(1) $\gamma[T[.,i,e]] \leq 1$ for all $(i,e) \in I \times E$

(2) for every $p \in S_M$, there exists $n \geq 1$ and a sequence

$$\{(i_0,e_0), (i_1,e_1) \cdots (i_n,e_n)\} \in (I \times E)^{n+1} \text{ such that}$$

$$\gamma [T^* [., \{(i_0,e_0), (i_1,e_1) \cdots (i_n,e_n)\}]] < 1$$

and

$$\text{Prob } [(i(\ell) = i_\ell, \ e(\ell) = e_\ell) : 0 < \ell < n \mid p(0) = p] > 0.$$

Remark 1.2: It is easily seen that if

$$\gamma[T[., i,e]] < 1 \text{ for all } (i,e) \in I \times E \tag{1.15}$$

then the algorithm is distance diminishing. The now classic Bush-Mosteller's algorithm, as an example, satisfies this criterion. In fact (1.15) is a very useful condition since it can be easily verified in many applications. For a more general definition of distance diminishing refer Norman [N10]

Definition 1.4: A state $p \in S_M$ is called an <u>absorbing state</u> of the algorithm T if

$$T [p, i,e] = p \tag{1.16}$$

for all (i,e) $I \times E$ (with probability one). A learning algorithm is said to be absorbing if and only if there is at least one absorbing state.

For reasons that will become clear (in chapter 3), in the context of absorbing algorithms our interest is in the class of algorithms for which V_M is the only set of absorbing states. We call such an algorithm an <u>absorbing barrier</u> algorithm. A <u>non absorbing barrier</u> algorithm is one for which there is no absorbing state.

Remark 1.3: In the case of absorbing barrier algorithm, the behavior of $p(k)$
for large k is very much a function of $p(o)$. However, the asymptotic
behavior of $p(k)$ in the case of non-absorbing barrier algorithm is independent of $p(o)$. In fact it will be shown in Chapter 2 that the distribution
of $p(k)$ after suitable normalization tends to be a normal distribution.
Henceforth, we shall call the non-absorbing barrier learning algorithms
Ergodic-algorithms.

1.5. Organization of the Book:

    This book is divided into two parts. Part I deals with
the theory and analysis of learning algorithms. Excluding this introduc-
tory chapter, part I contains Chapters 2, 3 and 4. In Chapter 2 a general
class of Ergodic learning algorithms are analyzed using the theory of
Markov processes that evolve in small steps. This class as a special
case includes the well known linear reward-penalty algorithm for experimenter-
subject controlled models in mathematical psychology [B10] It is well known
[C1] that this latter algorithm is only expedient. By proper choice of
certain functions and parameters in the learning algorithms, it is shown that
Ergodic algorithms are indeed ε-optimal. It may be mentioned that a system-
atic study of ε-optimal Ergodic algorithms is relatively new [L8] and some of
the results in this chapter appear for the first time.

Absolutely expedient algorithms [L2] and the analysis of the associated Markov process are the topics of Chapter 3. Till recently these are the only subclass of the absorbing barrier algorithms that are well understood. It is shown that by a proper choice of parameters of the learning algorithm, the probability of choosing the coin with maximum probability of success can be made as close to unity as desired. As a corollary it follows that absolute expediency implies ε-optimality. Almost all the learning algorithms that are known to date can be obtained as a special case of the algorithms in Chapters 2 and 3.

In Chapter 4 we explore the relationship between the learning algorithms of the type (1.3) and general stochastic approximation algorithms. In particular, by letting the step-size parameter vary with k, that is, by replacing θ by $\theta(k) \in (0,1]$ it is shown that a variety of stochastic approximation like algorithms results. Under suitable conditions on $\theta(k)$, stronger versions of the convergence results presented in Chapter 2 are obtained. As an example, in addition to all the conditions noted in Chapter 2, if

(1) $\sum_k \theta(k) = \infty$ and $\sum_k \theta^2(k) < \infty$, then it is shown that $\lim_{k \to \infty} \eta(k) = \eta^*$ with probability one where η^* can be made as close to $d_1 (= \max_i \{d_i\})$ as desired.

(2) $\sum_k \theta(k) = \infty$ and $\theta(k) \to 0$, then it is shown that $\lim_{k \to \infty} \eta(k) = \eta^*$ in probability where η^* is the same as in (1) above. Our aim is to highlight the different techniques that are available for the analysis of time-varying learning algorithms. In addition to the now classical method [W1] we also present a new class of techniques called "Compactness Method". This method is due to Kushner [K11] and for a detailed exposition of this latter method and its applications we refer the reader to a recent monograph by Kushner and Clark [K10]

Part II contains three different applications of learning algorithms developed in Part I. In Chapter 5 & 6 we present an interesting analysis of the two-person zero sum stochastic games with incomplete and imperfect information [H1] [S6] in which the players use the learning algorithms developed in part I. In particular we consider the situation in which neither player has no knowledge of the game matrix or the set of pure strategies available to the other player or the pure strategy actually chosen by the other player at any stage of the game. Based on just the random outcome (which is either a success or failure and whose distribution, as a function of the pure strategies available to either players, is also unknown to both players) the players update their mixed strategies using a learning algorithm. Our principal results of Chapters 5 and 6 may be summarized respectively as follows.

(1) If the game matrix has a saddle-point in pure strategies and if both players use an absolutely expedient algorithm of the linear reward-inaction type, then by proper choice of certain parameters of the algorithm either player can obtain an expected pay-off which is as close to Von-Neumann value of the game as desired.

(2) If the players use an Ergodic algorithm of the linear reward-penalty type with penalty term very small compared to reward term (in magnitude) then by proper choice of various parameters, either players can obtain an expected payoff as close to the Von-Neumann value as desired irrespective of whether or not the underlying game matrix has a saddle-point in pure strategies. A variety of simulation results are presented to further illustrate the theory.

Chapter 7 deals with a problem which has all the characteristics of the one treated in Chapter 5 except that in this case it is not a zero sum game but a perfect cooperative game in the sense that any outcome (success or failure) is equally shared by both players. Such a game is also known as "games with common fund" in which all players after each play pool together all their individual outcomes and share them equally [T2] . It is shown that if the players use a class of Ergodic learning algorithm, then their expected payoff can be made as close to maximum payoff as desired. It should be interesting to note that the above problem is also the two-person team problem [M2] with no information transfer and with no-apriori information. Thus results of this chapter may be viewed as extensions of some of those presented in Marshak and Radner [M2].

Control of a Markov chain when both the transition dynamics and reward structure are not known is considered in Chapter 8. Under the only assumption that the state of the Markov chain is known at each instant, by associating a learning algorithm of the absolutely expedient type with each state of the Markov chain, it is shown that there exists a proper choice of parameters of the learning algorithm such that the ONE STEP OPTIMAL CONTROL VECTOR can be obtained with a probability as close to unity as desired. Extensions of these results to the case when the state of the Markov chain is exactly known but with unit delay are also given. In fact, a more general learning algorithm of the absolutely expedient type along with a variety of simulation results are also presented. The results of this chapter can be regarded as an application of learning theory to the problems of adaptive control which bypasses the need for explicit identification [N5].

It must be mentioned that the Chapters 5, 6, 7 & 8 are essentially independent and can be read in any order after an initial familiarity with the first three chapters. There have been a number of other applications of these learning algorithms [N6] [T2] [N1] and the above choice only reflects the interest of the author.

1.6. Comments and Historical Remarks:

Our aim in this section is not to write or reconstruct the history of these learning algorithms but to refer the reader to the related literature.

If $M = 2$, that is, when there are only two coins, the problem reduces to the well-known two-armed bandit problem which has been studied rather extensively by mathematicians and statisticians in the U.S.A: Bellman [B3], Robbins [R1] [R2] Isbell [I3], Smith and Pyke[S11], Samuels [S1] Cover [C7], Cover and Hellman [C6], Sobel and Weiss[S12], Feldman [F1], to mention a few. The problem as we have stated and the concept of expediency (for $M \geq 2$) to our knowledge was first introduced by Tsetlin in 1961 [T3]. The work of Tsetlin in U.S.S.R. was later followed by various authors: Krylov [K4], Krinskii [K3], Ponomarev [P2]. It is rather surprising to note that in spite of the independent efforts much of the algorithms that were developed by these two groups of authors are strikingly similar. As an example, the algorithms of Robbins [R1] and Tsetlin [T3] are exactly the same (when $M = 2$) and is called the play the winner rule. The algorithm of Robbins [R2] and Krinskii [K3] are identical as also the algorithms of Smith and Pyke[S11] and Ponomarev [P2]. Both Samuels [S1] and Krylov [K4] deal with randomized transition rules in the algorithms. Mathematically all these above said algorithms reduce to the analysis of a finite state Markov chain which has a single Ergodic class. A recent survey paper by Witten [W3] discusses various aspects of the two-armed bandit problem.

Algorithms of the type given in (1.3) were originally introduced in Systems Theory by Varshavskii and Vorontsova [V2] and were called variable

structure learning algorithms. This work was later followed by Fu [F7],
McMurtry and Fu [M4], Mclaren [M3], Chandrasekaran and Shen [C1], Shapiro
and Narendra [S7], Viswanathan and Narendra [V3], Lakshmivarahan and
Thathachar [L1] [L2], Herkenrath and Theodorescu [H2], Baba and Sawaragi [B1],
Witten [W2] and Aso and Kimura [A3]. All these algorithms have their roots
in mathematical psychology, especially in the works of Bush and Mosteller
[B10], Luce [L14] and Norman [N10]. In spite of the differences in their
origin, much of the questions that arise in the mathematical analysis
of these algorithms are very similar. Recent work by Norman [N10] provides
the much needed unified mathematical framework for the analysis and design
of these algorithms and we make heavy use of the results from Norman.

The concept of expediency was originally introduced by Tsetlin. However, Tsetlin defined it in the context of deterministic algorithm (play
the winner rule as $\lim_{k \to \infty} \eta(k) > \eta_o$). Notice this is a very strong require-
ment compared to (1.7). The definition of expediency as given in (1.7)
is new and the concept of absolute expediency was introduced by Lakshmivarahan
and Thathachar [L2]. The definition of expediency and ε-optimality given
in (1.7) and (1.9) are generalization of the widely used ones, [L2] [N1].
A casual look at the earlier papers on learning algorithm may give the
impression to a novice that a number of learning algorithms are optimal.
Kushner, et al [K8] [V6] have shown by a counter example that most of the
so-called "optimal" algorithms are not optimal. At present there are a
variety of algorithms that are ε-optimal [L7] [L5] [V6] [B1].

Most of the materials in section 1.4 follow Norman [N10] and Iosifescu
and Theodorescu[I2]. Very recently Herkenrath, Kalin and Lakshmivarahan [H3]
have derived necessary and sufficient conditions for the general class of
algorithms of type (1.3) to be an absorbing barrier algorithm.

Almost all the papers on learning algorithms that have appeared in engineering literature have been couched in the language of learning automata. [V2] [M3] [M4] [C1] [S7] [V3] [L1] [L2] [B1] [W2] [H2] [N1] [T6] A Learning Automaton (LA) is a combination of finite-state stochastic automaton and a random-environment connected in a feed-back loop where the input to one is the output of the other. See the figure below:

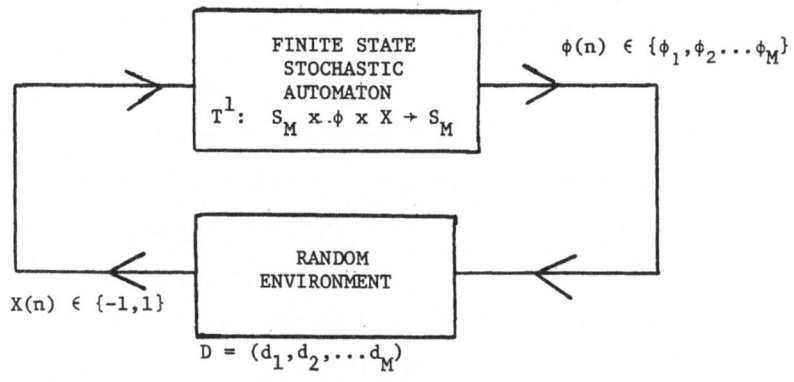

LEARNING AUTOMATON

For our purposes we can represent a LA as a quintuple $(X, \phi, \underline{p}, T', D)$ where

1) $x = \{-1, 1\}$ is the input set to the Stochastic automaton (same as output of the environment).

    If $x(k)$ is the input at time k, then $x(k) = 1$ is called a success and $x(k) = -1$ is called failure.

2) $\phi = \{\phi_1, \phi_2, \ldots \phi_M\}$ is the set of states of the stochastic automaton $(2 \leq M < \infty)$ without loss of generality let us assume that the state itself forms the output of the stochastic automaton (which is also the input to the random environment).

3) If $\phi(k)$ is the state of the stochastic automaton at time k, then $p(k) = (p_1(k), p_2(k), \ldots p_M(k))^T$ is the state probability distribution at time k where

   $$p_i(k) = \text{Prob} [\phi(k) = \phi_i].$$

4) $D = (d_1, d_2, \ldots d_M)^T$ is the success probability vector where

$\quad d_i = \text{Prob } [x(k) = +1 \mid \phi(k) = \phi_i]$

$\quad c_i = \text{Prob } [x(k) = -1 \mid \phi(k) = \phi_i] = 1 - d_i$

5) $T': \quad S_M \times \phi \times X \to S_M$ and

$\quad p(k + 1) = T'[p(k), \phi(k), x(k)]$

is a learning algorithm of the type (1.3).

In this setup the stochastic automaton is usually the decision device and the random environment is an abstract or mathematical model of a real-world problem. For the learning problem to be non-trivial it is assumed that the input-output characteristics of the random environment is unknown. In the context of the coin tossing experiment, the stochastic automaton being in state $\phi(k) = \phi_i$ reflects our decision to toss the ith coin at time instant k and $x(k) = +1$ or -1 represent the result of the toss being a head (success) or tail (failure) respectively.

In the above learning automaton set up (motivated by the coin-tossing learning problem) we have only considered random environments in which the output is a binary (two-valued) random variable. It is conceivable that output of the random-environment may take $N(2 < N < \infty)$ values or even have a continuous distribution. Learning Automata incorporating the above variations have been studied in the literature and are called Q and S-model learning automata [N1] [V3] [L3] [C4] [C1] [J1] [M3] [M4] [S7]. Most of the S-model learning automata were developed in the context of parameter self-optimisation with multimodal performance criteria [C4] [S7] [M3] [M4] [V3] [N1] We indicate these extensions at appropriate time in the body of the manuscript (mostly as exercises). It should be evident from the above discussion that given a learning algorithm of the type (1.3) or its possible variant, we can always associate a learning

automata with it and vice-versa. Having seen this, in this manuscript, we
will only talk about learning algorithms

There is also an intimate connection between the learning problem we
have described in the main body of this chapter and sequential sampling
techniques. For, our learning problem can be restated as follows; given
a class of M – binomial populations (each with unknown mean) develop a
method to make observations on samples from these populations so as to
determine which particular population has the largest mean value. The
decision as to which population is to sample next must depend on the results
of the previous observations. Notice instead of binomial if we allow
arbitrary discrete distributions for each population we get the so called
Q-model learning automata models. We let the distribution of each population to be continuous, the S-model learning automata results. We refer
the reader to the papers by Sobel and Weiss [S12], Fox [F3] Bradt, Johnson
and Karlin [B8] and a monograph by Bechhofer, Kiefer and Sobel [B2] for
various discussions on sequential sampling techniques and other related
topics.

1.7 Exercises

1.1) A subset F of S_M is said to be __stochastically closed__ if and only if

for every s $\in$ F,

T[s,i,e] $\in$ F with probability 1. Furthur, a subset F is called

an __ergodic kernel__ if F is stochastically and topologically closed

and there is no stochastically and topologically closed proper subset

of F.

(a) Show that two distinct ergodic kernels are disjoint.
(b) Show that an __absorbing state__ is an ergodic kernel.

(c) Find conditions on the functions $f_s^i(.)$ and $g_s^i(.) i,s = 1,2,...,M$

figuring in the algorithm (1.3) such that

(1) S_M is the only one ergodic kernel.

(2) each element of V_M is an ergodic kernel

1.2) Let $\delta p_i(k) \triangleq p_i(k+1) - p_i(k)$. Define

$$VAR(p_i) = E\{ [\delta p_i(k)]^2 \mid p(k) = p] - E^2[\delta p_i(k) \mid p(k) = p]$$

Notice $VAR(p_i)$ is the conditional variance of the one step increment in

$p_i(k)$ when $p(k) = p$. Show that, if VAR $(p_i) > 0$ for all $P_i \in [0,1]$ and

for all i = 1,2,...M, then S_M is the only one ergodic kernel.

1.3) Find (sufficient) conditions for the algorithms (1.3) to be an ergodic

algorithm.

1.4) Find an example of the functions $f_s^i(.)$ and $g_s^i(.) i,s = 1,2,...,M$ such

that the algorithm (1.3) is __distance diminishing__.

CHAPTER 2

ERGODIC LEARNING ALGORITHMS

2.1. Introduction:

This chapter presents an analysis of <u>general non-linear reward-penalty</u> <u>ergodic-N_{R-P}^{E} algorithms</u>. The basic property that characterizes this class of algorithms is that all the states under this class of algorithms are <u>non-absorbing</u>. The now classic linear reward-penalty - L_{R-P}^{E} algorithm is a special case of this algorithm. It is well known [C1] that this L_{R-P}^{E} algorithm is only expedient. Using the theory of Markov processes that evolve by small steps [N14] a variety of characterizations of the process $\{p(k)\}$ $k \geq 0$ such as the evolution of the mean and variance and in fact its actual sample path behavior are given. As a by-product, it is proved that there exists a proper choice of parameters and functions such that the N_{R-P}^{E} algorithm is ε-optimal.

In order to bring out the key ideas of the approach we first present an analysis of the case when M = 2 (that is, when there are only two coins) and then of the general case M $\geq$ 2. A number of examples along with the results of simulation experiments are given to further illustrate the properties of these algorithms. The study of ε-optimal ergodic algorithms is relatively new and most of our presentation follows [L7].

Our notations and conventions in this and later chapters follow those set out in Chapter 1.

2.2. N_{R-P}^{E} - ALGORITHM

Recall that $p_s(k)$ is the probability of selecting the coin s at time k where $p(k) = (p_1(k), p_2(k) \ldots, p_M(k))^T$. Suppose the coin i (=1,2,...,M) was chosen as a sample realization from $p(k)$. Then $p(k+1)$ is defined as follows[1]

$$
\begin{aligned}
p_i(k+1) &= p_i(k) + \theta\lambda[p(k)]\,[1-p_i(k)] \\
p_j(k+1) &= p_j(k) - \theta\lambda[p(k)]\,p_j(k), \; j \neq i
\end{aligned}
\left.\right\} \quad \text{if toss of coin i resulted in } \underline{\text{success}}
$$

$$
\begin{aligned}
p_i(k+1) &= p_i(k) - \theta\psi_i[p(k)] \\
p_j(k+1) &= p_j(k) + \frac{\theta}{(M-1)}\,\psi_i[p(k)], \; j \neq i
\end{aligned}
\left.\right\} \quad \text{if toss of coin i resulted in } \underline{\text{failure}}
$$

(2.1)

where $\theta \in (0,1]$ and $\lambda : S_M \to [0,1]$ and $\psi_s : S_M \to [0,1]$ s = 1, 2,...,M such that

(C2.1) Either $\lambda[p] \equiv 0$ or $0 \leq \lambda[p] < 1 \; \forall \; p \in S_M$

with equality holding only if $p \in V_M$

(C2.2) $0 \leq \psi_s[p] < p_s \; \forall \; p_s \in [0,1]$, s = 1,2,...,M

with equality holding only at $p_s = 0$ and

(C2.3) $\lambda[.]$ and $\psi_s[.]$ are twice continuously differentiable functions on S_M. Conditions (C2.1) - (C2.3) ensure that $p(k) \in S_M$ for all k if $p(0)$ does. Clearly $\{p^{\theta}(k)\}$ is an indexed family of stationary Markov processes and θ essentially controls the step size of the evolution of the above Markov process. In the following for notational simplicity we shall drop the superscript θ.

Typical choice of functions satisfying (C2.1) - (C2.3) and that are of interest to us are:

(a) $\lambda[p] = a_1 + a_2\, p_1^{r_1} p_2^{r_2} \cdots p_M^{r_M}$ where $a_1 \geq 0$, $a_2 \geq 0$ and r_i's are non-negative integers

(b) $\psi_i[p] = b_1^i\, p_1^{m_1} p_2^{m_2} \cdots p_M^{m_M} + b_2^i\, p_i\, [b_3^i - b_4^i\, p_i]$

where m_j's are non-negative integers and $0 \leq b_1^i$ and $0 < b_2^i, b_3^i, b_4^i < 1$, $b_3^i > b_4^i$

---

[1] The increase (decrease) in $p_j(k)$ when the toss of coin i resulted in success (failure) is called Reward (Penalty).

for all $i = 1, 2, \ldots, M$. Note m_j's also may depend on i.

As a first observation we have the following

Lemma 2.1: Conditions (C2.1) - (C2.3) imply that the algorithm (2.1) is non-absorbing.

Proof: If $p_i(k) = r \in (0,1]$, then it follows from (2.1) that

$$p_i(k+1) = p_i(k) - \theta \, \psi_i[p(k)] < p_i(k)$$

with probability $c_i > 0$. If $p_i(k) = 0$, then there exists some $j \neq i$ for which $p_j(k) \in (0,1]$ and by the same argument $p_j(k+1) < p_j(k)$ with probability $c_j > 0$. Hence, for all possible values of $p_i(k)$, $p(k+1) \neq p(k)$ with positive probability. As i is arbitrary, the lemma follows.   Q.E.D.

Define

$$\delta p(k) \triangleq (\delta p_1(k), \, \delta p_2(k), \, \ldots, \, \delta p_M(k) \,)^T$$

where $\delta p_i(k) = p_i(k+1) - p_i(k)$. To see what we should expect in this chapter, let $p(0) = p$ with probability one. Consider

$$p(k) - E[p(k)] = \sum_{m=0}^{k-1} \{\delta p(m) - E[\delta p(m)]\}$$

Let

$$y_m \triangleq \delta p(m) / \theta$$

$$z(k) \triangleq \theta^{-1/2} [p(k) - E[p(k)] \,]$$

$$= (k\theta)^{1/2} \sum_{m=0}^{k-1} [y_m - E[y_m] \,] k^{-1/2}$$

Assume for a moment that the distribution of $\delta p(m)$ does not depend on $p(m) = p$ for all $m \geq 0$. As $\{p(k)\}$ is a Markov process, this assumption implies that $\{y_m\}$ is a sequence of independent, identically distributed random vectors and by central limit theorem [D2] as $\theta \to 0$, $k\theta \to t > 0$, the distribution of z_k for large k converges to the normal distribution $N [0, t \, \sigma^2]$ with mean zero and covariance matrix $t\sigma^2$ where σ^2 is the covariance matrix of y_m. From algorithm (2.1) we do know the distribution $\delta p(m)$ depends on p. However as

y_m's are still independent, we can expect a central limit theorem to hold for our process $\{p(k)\}$.

In the following we shall collectively record some of the properties of $\delta p(k)$ for our later use. From (2.1) it is easily seen that

(P2.1) $\qquad E[\delta p(k) \mid p(k) = p] \underset{=}{\Delta} \theta \underline{W}(p)$

where $\underline{W}(p) = (W_1(p), W_2(p),\ldots, W_M(p))^T$

$\qquad W_i(p) = W_i^R(p) + W_i^P(p)$, $i = 1,2,\ldots,M$

$\qquad W_i^R(p) = \lambda[p] \; p_i \sum_{j \neq i} p_j \; [d_i - d_j]$

$\qquad W_i^P(p) = \dfrac{1}{(M-1)} \sum_{j \neq 1} [p_j \; \psi_j [p] \; c_j - p_i \; \psi_i [p] \; c_i]$

and $d_s = 1 - c_s$, $s = 1, 2, \ldots, M$.

As $p(k) \; \epsilon \; S_M$ for all $k \geq 0$, it follows that

$$\sum_{i=1}^{M} W_i(p) = 0 \qquad \forall \; p \; \epsilon \; S_M \tag{2.2}$$

The superscripts R and P of $W_i^R(.)$ and $W_i^P(.)$ refer to the part $W_i(.)$ that results from reward and penalty terms of the algorithm (2.1) respectively. Similarly

(P2.2) $\qquad E[\;\delta p(k) \; \delta p^T(k) \mid p(k) = p] \underset{=}{\Delta} \theta^2 \; a(p)$

and

$\qquad s(p) \underset{=}{\Delta} a(p) - \underline{W}(p) \; \underline{W}^T(p)$

where

$\qquad a_{ij}(p) = a_{ij}^R(p) + a_{ij}^P(p)$

$a_{ii}^R(p) = \lambda^2[p] \; \{(1-p_i)^2 \; p_i \; d_i + p_i^2 \sum_{j \neq i} p_j \; d_j\} > 0, \forall \; p \not\in V_M$

$a_{ii}^P(p) = \{\psi_i^2[p] \cdot p_i \; c_i + \dfrac{1}{(M-1)^2} \sum_{j \neq i} \psi_j^2[p] \; p_j \; c_j\} > 0, \quad \forall \; p \; \epsilon \; S_M$

$$a_{ij}^R(p) = -\lambda^2[p] \{p_i(1-p_i) \; p_j \; d_i + \sum_{j \neq \boldsymbol{i}} p_j \; p_i \; (1-p_j) \; d_j\} < 0, \; \forall \; p \notin V_M$$

$$a_{ij}^P(p) = -\frac{1}{(m-1)} \{\sum_{s=1}^{M} p_s \; \psi_s^2 \; [p] \; c_s\} < 0, \; \forall \; p \in S_M.$$

The following properties P2.3 - P2.5 are easily checked by routine arguments and we omit the details.

(P2.3) $\qquad E[\; |\; \delta p(k) \; |^3 \; | \; p(k) = p] = 0(\theta^3)$

(2)

where $|x|$ refers to the Eulidean norm of the vector x and the order of magnitude is uniform in p and k.

(P2.4) $\underline{W}(p)$ is twice continuously differentiable in S_M

(P2.5) s(p) is differentiable in S_M.

Remark 2.1: (P2.4) and (P2.5) easily follow from (C2.3) and (P2.1) and (P2.2) and we explicitly state them here for later reference. If $A = \{a_{ij}\}$ is an M x M matrix, let

$$|A|^2 = \sum_{i,j=1}^{M} a_{ij}^2$$

As S_M is compact, (P2.4) and (P2.5) imply $W_i(p)$ and $a_{ij}(p)$ are bounded functions in S_M. Thus $|\underline{W}(p)|$ and $|s(p)|$ are bounded in S_M. Using this observation, from P2.2 we have

$$E \; \{ \; |[\delta p(k) - \underline{W}(p)] \; [\delta p(k) - \underline{W}(p)]^T | \; | \; p(k) = p \} = 0(\theta^2) \qquad (2.3)$$

2.3 Analysis:

In order to bring out the key ideas of the approach, in this section, let us confine our attention to the case when M = 2. Let us denote $\mathbf{p} = (P, 1-P)^T$ and with little abuse of notation redefine $W(P) \triangleq W_1(p)$, $a(\mathbf{P}) \triangleq a_{11}(p)$, $\lambda[\mathbf{P}] \triangleq \lambda[p]$ and $\psi_j[\mathbf{P}] \triangleq \psi_j[p]$. Let us rewrite (2.1) using this new notation:

_____
(2)
$\quad 0(\theta^n) = \sum_{i=0}^{n} k_i \; \theta^i$ where k_i are all positive constants.

$$P(k+1) = \begin{cases} P(k) + \theta \lambda[P(k)] [1 - P(k)] & \text{if coin 1 tossed and success} \quad (3) \\ P(k) - \theta \psi_1 [P(k)] & \text{if coin 1 tossed and failure} \\ P(k) - \theta \lambda[P(k)] P(k) & \text{if coin 2 tossed and success} \\ P(k) + \theta \psi_2[P(k)] & \text{if coin 2 tossed and failure} \end{cases}$$

(2.1a)

Clearly $\{P(k)\}$ is a stationary Markov process in $[0,1]$. From P2.1 and P2.2 we have

$$\left. \begin{aligned} W(P) &= W^R(P) + W^P(P) \\ W^R(P) &= \lambda[P] \, P \, (1-P) \, (d_1 - d_2) \\ W^P(P) &= [(1-P) \, \psi_2 \, [P] \, \mathbf{c}_2 - P \, \psi_1 \, [P] \, \mathbf{c}_1] \end{aligned} \right\} \qquad (2.4)$$

and

$$\left. \begin{aligned} s(P) &= a(P) - W^2(P) \\ a(P) &= a^R(P) + a^P(P) \\ a^R(P) &= P(1-P) \, \lambda^2[P] \, [(1-P) \, d_1 + P \, d_2] \\ a^P(P) &= \{(1-P) \, \mathbf{c}_2 \, \psi_2^{\,2} \, [P] + P \, \mathbf{c}_1 \, \psi_1^{\,2} \, [P] \} \end{aligned} \right\} \qquad (2.5)$$

The following theorem characterizes the zeros of the function $W(P)$.

Theorem 2.1: In addition to (C2.1) - (C2.3) if

$$(1-P) \, \psi_2 \, [P] \gtreqless P \, \psi_1[P] \text{ for } P \lesseqgtr \frac{1}{2} \qquad (2.6)$$

then there exists unique β_1 and β_2 in $(0,1)$ such that $W^P(\beta_1) = 0$ and $W(\beta_2) = 0$ and $\beta_2 \gtreqless \beta_1 \gtreqless \frac{1}{2}$ according as $\mathbf{c}_2 \gtreqless \mathbf{c}_1$. (2.7)

Proof: Let $\mathbf{c}_2 - \mathbf{c}_1 = d_1 - d_2 = \delta$ and rewrite $W^P(P)$ as

$$W^P(P) = \mathbf{c}_1 \, [(1-P) \, \psi_2 \, [P] - P \, \psi_1 \, [P]] + (1-P) \, \psi_2 \, [P] \, \delta \qquad (2.8)$$

If $\delta = 0$, $W^P(0) = \mathbf{c}_1 \, \psi_2 \, (0) > 0$ and $W^P(1) = -\mathbf{c}_1 \, \psi_1 \, (1) < 0$ and from (2.6) it follows that $1/2$ is the unique point in $[0,1]$ such that $W^P(1/2) = 0$. If

---

[3] P as the argument of a function refers to the vector $(P, 1-P)^T$, otherwise denotes a scalar.

$\delta > 0$, then $W^P(1/2) > 0$ and $W^P(1) = -\psi_1[1] \mathbf{c}_1 < 0$. Thus the effect of $\delta > 0$ is to translate the zero of $W^P(P)$ to the right. Hence, there exists an unique $\beta_1 > 1/2$ such that $W^P(\beta_1) = 0$.

If $\delta > 0$, then $W(\beta_1) = W^R(\beta_1) > 0$ and

$$W(1) = -\psi_1[1] \mathbf{c}_1 < 0.$$

Hence, once again the effect of adding $W^R(P)$ to $W^P(P)$ is to further translate the zero of $W(P)$ to the right of β_1. Thus there exists unique $\beta_2 > \beta_1$ such that $W(\beta_2) = 0$. Similar arguments hold when $\delta < 0$. (Q.E.D.)

Example 2.1: Let $\lambda(P) = .1 + .1 P(1-P)$

$$\psi_1(P) = P [.55 - .5P]$$

$$\psi_2(P) = (1-P) [.55 - .5(1-P)]$$

$W^P(P)$, $W^R(P)$ and $W(P)$ for various values of P are tabulated below.

Table 2.1

$\mathbf{c}_1 = 0.4$, $\mathbf{c}_2 = 0.65$
$\beta_1 = 0.6025$ $\beta_2 = 0.64320$

P	$W^P(P)$	$W^R(P)$	$W(P)$
0.0	0.03250	0.0	0.03250
0.1	0.05065	0.00245	0.05310
0.2	0.05520	0.00464	0.05984
0.3	0.04930	0.00635	0.05565
0.4	0.03610	0.00744	0.04354
0.5	0.01875	0.00781	0.02656
0.6	0.00400	0.00744	0.00784
0.7	-0.01580	0.00635	-0.00945
0.8	-0.02670	0.00464	-0.02206
0.9	-0.02915	0.00245	-0.02670
1.0	-0.02000	0.0	-0.02000

In the algorithm (2.1a) if we replace $\psi_j[P]$ by $\alpha\psi_j[P]$ where α is real and positive, it is easily seen that $W^P[P]$ and $a^P(P)$ in (2.4) and (2.5) get multiplied by α and α^2 respectively.

Define

$$W(\alpha,P) \triangleq W^R(P) + \alpha W^P(P) \tag{2.9}$$

where α is real and $0 < \alpha \leq 1$ where $W(1,P) = W(P)$.

Corollary 2.1: If all the conditions of theorem 2.1 hold, then there exists an unique $\beta(\alpha) \in (0,1)$ such that $W(\alpha, \beta(\alpha)) = 0$ and

   a) $\beta(\alpha) \geq \beta_2$ and $\beta(\alpha) \to 1$ as $\alpha \to 0$ when $\mathbf{c}_2 > \mathbf{c}_1$

and

   b) $\beta(\alpha) \leq \beta_2$ and $\beta(\alpha) \to 0$ as $\alpha \to 0$ when $\mathbf{c}_2 < \mathbf{c}_1$

Proof: On rewriting (2.9) becomes

$$W(\alpha,P) = W(1,P) - (1 - \alpha) W^P(P) \tag{2.10}$$

where $W(1,P) = W(P)$. From the definition of β_2 and the properties of $W^P(P)$ we have

$$W[\alpha,\beta_2] = - (1 - \alpha) W^P(\beta_2) > 0 \qquad \text{and}$$

$$W[\alpha,1] = -\alpha\mathbf{c}_1 \psi_1[1] < 0.$$

Thus, the effect of multiplying $W^P(P)$ is to still further shift the zero crossing point $\beta(\alpha)$ to the right of β_2 when $\mathbf{c}_2 > \mathbf{c}_1$. That is, $W[\alpha, \beta(\alpha)] = 0$ and $\beta(\alpha) \geq \beta_2$. If $\alpha' < \alpha$, then it can be seen that

$$W[\alpha', \beta(\alpha)] = W^P(\beta(\alpha)) (\alpha' - \alpha) > 0 \tag{2.11}$$

where we have used the definition of $\beta(\alpha)$ namely $W(\alpha,\beta(\alpha)) = 0$. This implies $\beta(\alpha)$ increases as α decreases. Further as $W(0,1) = 0$, it follows that 1 is the least upper bound on $\beta(\alpha)$ and part (a) of corollary is proved. Part (b) follows by similar arguments.   Q.E.D.

Example 2.1 (continued).

The variation of $\beta(\alpha)$ with α is given in table 2.2.

Table 2.2

$$C_1 = 0.40 \qquad C_2 = 0.65$$

α	$\beta(\alpha)$
1.0	0.64324
0.9	0.64768
0.8	0.65318
0.7	0.66019
0.6	0.66941
0.5	0.68207
0.4	0.70046
0.3	0.72938
0.2	0.78020
0.1	0.87867
0.05	0.94668
0.02	0.98186
0.01	0.99146

Let $W'(\alpha,P)$ represent the derivative of $W(\alpha,P)$ with respect to P. As a consequence of the above corollary 2.1 we have the following:

Corollary 2.2: $W'[\alpha,\beta(\alpha)] < 0$ for all $\alpha \epsilon$ (0,1]

Proof:

From Corollary 2.1 we have

$$W(\alpha,P) \gtreqless 0 \quad \text{for } P \lesseqgtr \beta(\alpha) \tag{2.12}$$

From (P2.4) and (2.12) we easily see that

$$\lim_{\epsilon \to 0} \frac{W(\alpha, \ \beta(\alpha) + \epsilon) - W(\alpha,\beta(\alpha) - \epsilon)}{2\epsilon}$$

exists and $w'(\alpha,\beta(\alpha)) < 0$ \hfill Q.E.D.

Now that we have seen the effect of multiplying the functions $\psi_i(p)$ in the algorithm (2.1a) by the factor α, in the following unless stated otherwise we shall only consider the case when $\alpha = 1$. In other words, we will only consider the function $W(P)$ instead of $W(\alpha, P)$. However, as will be evident, for every theorem we have using $W(P)$ there is a corresponding one with $W(\alpha, P)$ where $0 < \alpha \leq 1$.

Example 2.1: (Continued) The values of $W'(\alpha, \beta(\alpha))$ and $s[\beta(\alpha)]$ for various values of α are given table 2.3. Information from this table 2.3 will be later used to compute the asymptotic variance of $P(k)$ after suitable normalization.

Table 2.3

$$C_1 = 0.4 \qquad C_2 = 0.65$$

α	$W'[\alpha, \beta(\alpha)]$	$s[\beta(\alpha)]$	$\dfrac{s[\beta(\alpha)]}{2\lvert W'[\alpha, \beta(\alpha)]\rvert}$
1.0	-0.17621	1.11520×10^{-2}	0.03164
0.9	-0.15851	0.92285×10^{-2}	0.02911
0.8	-0.14082	0.75102×10^{-2}	0.02667
0.7	-0.12315	0.59955×10^{-2}	0.02434
0.6	-0.10551	0.46816×10^{-2}	0.02219
0.5	-0.87926×10^{-1}	0.35632×10^{-2}	0.02026
0.4	-0.70432×10^{-1}	0.26304×10^{-2}	0.01867
0.3	-0.53143×10^{-1}	0.18610×10^{-2}	0.01751
0.2	-0.36438×10^{-1}	0.12008×10^{-2}	0.01648
0.01	-0.22548×10^{-1}	0.53255×10^{-3}	0.01181
0.05	-0.20297×10^{-1}	0.20745×10^{-3}	0.00511
0.02	-0.22270×10^{-1}	0.65953×10^{-4}	0.00148
0.01	-0.23514×10^{-1}	0.30453×10^{-4}	0.00065

In the following we first explore the consequences of the Theorem 2.1 and its corollaries. Consider the following differential equation:

$$f'(t) = W(f(t)) \tag{2.13}$$

where $f(0) = P$. As $W(.)$ is twice differentiable (refer (P2.4)) solution to (2.13) exists and is unique.

Define

$$R(x,y) = \begin{cases} \dfrac{W(x) - W(y)}{x - y} & \text{if } x \neq y \\[2mm] w'(y) & \text{otherwise.} \end{cases} \tag{2.14a}$$

From Corollary 2.2 we have

$$R(x, \beta_2) < 0 \qquad \text{for all } x \in [0,1]$$

where recall that $W(\beta_2) = 0$.

Let

$$-\gamma = \sup_{x \in [0,1]} R(x, \beta_2) < 0 \tag{2.14b}$$

Clearly

$$\frac{d}{dt} (f(t) - \beta_2)^2 = 2[f(t) - \beta_2]^2 R(x, \beta_2)$$
$$\leq -2\gamma [f(t) - \beta_2]^2$$
$$< 0$$

Hence

$$|f(t) - \beta_2| \leq |P - \beta_2| e^{-\gamma t} \leq e^{-\gamma t} \tag{2.14c}$$

Thus, β_2 is the only uniformly asymptotically stable solution [L11] of (2.13) and $f(t) \rightarrow \beta_2$ as $t \rightarrow \infty$.

As

$$f''(t) = W'(f(t)) \ W(f(t))$$

from (P2.4) it follows that

$$|f''(t)| \leq |W'(f(t))| \quad |W(f(t))| < L \tag{2.15}$$

Where L is a constant. If we denote $f(k) \triangleq f(k\theta)$, using second order Taylor Series expansion, in view of (2.15), we have

$$\delta f(k) = f(k+1) - f(k)$$
$$= \theta W(f(k)) + 0(\theta^2) \tag{2.16}$$

uniformly for all $k \geq 0$ and $f(0) = P \epsilon [0,1]$

With these preliminary observations we now turn to the characterization of the properties of first two moments of the Markov process $\{P(k)\}$.

Theorem 2.2. In addition to (P2.1) - (P2.5) if the function W(P) has a unique zero β_2 in [0,1] and $W'(\beta_2) < 0$, then

$$E\{[P(k) - f(k)]^2 \mid P(0) = P\] = 0(\theta)$$

uniformly for all $k \geq 0$ and $P \epsilon [0,1]$.

Proof: Let $a_k = E\{[P(k) - f(k)]^2\}$

From this we have

$$a_{k+1} = a_k + 2 E\{[\delta P(k) - \delta f(k)][P(k) - f(k)]\} + E\{[\delta P(k) - \delta F(k)]^2\} \tag{2.17}$$

But as $(x - y)^2 \leq 2(x^2 + y^2)$ we have

$$E\{[\delta P(k) - \delta f(k)]^2\} \leq 2 E\{[\delta P(k)]^2\} + 2 E\{[\delta f(k)]^2\}$$
$$\leq 2 \theta^2 [a(P) + W^2(f(k))] + 0(\theta^3)$$
$$\leq L_1 \theta^2 = 0(\theta^2) \tag{2.18}$$

where the second inequality follows from (P2.2), (2.5) and (2.16), the third inequality follows from the fact that $a(p)$ and W(P) are bounded functions and here and elsewhere L and L_j, $j = 1,2,3,\ldots$, etc. are suitable positive constants.

Similarly

$$E\{[\delta P(k) - \delta f(k)] [P(k) - f(k)]\}$$

$$= E \{[E [\delta P(k)| P(k)] - \delta f(k)] [P(k) - f(k)]\}$$

$$= \theta E \{[W(P(k) - W(f(k))] [P(k) - f(k)]\} + O(\theta^2)$$

$$= \theta E \{[R(P(k),f(k))][P(k) - f(k)]^2\} + O(\theta^2) \qquad (2.19)$$

where the first equality follows from the iterated law of conditional expectation [D2], the second follows from (P2.1) and (2.16) and the third one
follows from the definition of $R(x,y)$.

The properties of $W(.)$ implies that $R(x,y)$ is uniformly continuous in
$[0,1]$ x $[0,1]$. This in turn implies that there exists a $\delta > 0$ (depending
only on γ) and small enough such that if $|y - \beta_2| \leq \delta$, then

$$|R(x,y) - R(x,\beta_2)| \leq \frac{\gamma}{2} \qquad (2.20)$$

for all $x \in [0,1]$. Hence

$$|R(x,y)| \leq \frac{\gamma}{2} \qquad (2.21)$$

for all $x \in [0,1]$ and $|y - \beta_2| \leq \delta$

Now given a $\delta > 0$ we can find a positive integer N large enough such that for
all $k\theta > N$

$$|f(k) - \beta_2| \leq e^{-\gamma N} < \delta \qquad (2.22)$$

From (2.21) and (2.22) it follows that for $k\theta > N$

$$R(P(k), f(k)) \leq - \frac{\gamma}{2} \qquad (2.23)$$

Combining (2.17) - (2.19) and (2.23) for $k\theta > N$ we have

$$a_{k+1} \leq a_k(1 - \gamma\theta) + L_2\theta^2 \qquad (2.24)$$

Another estimate which is generally valid is

$$R(\ P(k),\ f(k)\ ) \le \underset{(x,y)}{\text{Sup}}\ |R(x,y)|\ \underset{\triangle}{\underline{\;}}\ |R| \tag{2.25}$$

Using (2.25) in (2.17) - (2.19) we have

$$a_{k+1} \le a_k[1 + 2|R|\theta] + L_3\theta^2 \tag{2.26}$$

Iterating (2.26) and remembering $a_0 = 0$ (since $P(0) = P$ with probability one and $f(0) = P$)

$$a_k \le \frac{L_3}{2|R|}\ [\ (1 + 2|R|\theta\ )^k -1]\theta$$

$$\le \frac{L_4}{2|R|}\ [\ e^{2|R|k\theta}-1]\ \theta$$

If $k\theta \le N$, then we can write the above inequality as

$$a_k \le \frac{L_4}{2|R|}\ [\ e^{\ 2|R|N}-1]\ \theta \le L_5\ \theta \tag{2.27}$$

Iterating (2.24)

$$a_k \le (1 - \gamma\theta)^{k-i}\ a_i + \theta[1 - (1 - \gamma\theta)^{k-i}]\ \frac{L_2}{\gamma}$$

If $0 < \theta < \text{Min}(1,\ 1/\gamma)$, then

$$a_k \le a_i + L_3\theta \tag{2.28}$$

If $k\theta < N\ < i\theta$, then from (2.27) and (2.28)

$$a_k \le L_6\theta \text{ for all } k\ \theta\ >N \tag{2.29}$$

Combining (2.27) and (2.29)

$$E\{[\ P(k) - f(k)]^2\}\ \le L\theta$$

for all $P(0) = P \epsilon\ [0,1]$ and for all k.   Q.E.D.

Remark 2.2: A careful scrutiny of the proof reveals that the full force of (P2.1) - (P2.3) are not needed. All that is effectively required in the proof is that

$$E\{[\delta P(k)]^2 \mid P(k) = P\} = 0(\theta^2).$$

However, the properties of W(.) as given in P2.4 are well exploited.

Theorem 2.3: If all the conditions of Theorem 2.1 hold, then

$$E[P(k) \mid P(0) = P] = f(k) + 0(\theta)$$

uniformly for all $P \varepsilon [0,1]$ and $k \geq 0$.

Proof: From (P2.1)

$$E[P(k+1)] = E[P(k) + \theta W(P(k)]$$

By (2.16)

$$f(k+1) = f(k) + \theta W(F(k)) + L_1 \theta^2$$

Denoting $b_k \underset{=}{\triangle} E[P(k)] - f(k)$, it follows

$$|b_{k+1}| \leq |b_k| + \theta |[W(P(k)) - W(f(k))]| + L_2 \theta^2 \tag{2.29}$$

Since W(.) is twice differentiable

$$W(P(k)) - W(f(k)) = (P(k) - f(k)) W'(f(k)) + 0(|P(k) - f(k)|^2) \tag{2.30}$$

Substituting (2.30) into (2.29) we get

$$|b_{k+1}| \leq |b_k| [1 + \theta |W'(f(k))|] + \theta 0(|P(k) - f(k)|^2) + 0(\theta^2)$$

Taking expectations on both sides and using the estimate of the theorem 2.2,

the above inequality becomes

$$|b_{k+1}| = |b_k| [1 + \theta |W'(f(k))|]] + 0(\theta^2) \tag{2.31}$$

Conditions of the theorem 2.3 imply that

$$W'(\beta_2) \leq -\gamma < 0$$

For any given (small) $\delta > 0$, there exists a large positive integer N* such

that for all $k \theta > N^*$ we have

$$|f(k) - \beta_2| \leq e^{-N^* \gamma} < \delta \text{ and } |W'(f(k))| < \frac{\gamma}{2} \tag{2.32}$$

Also

$$W'(f(k)) \leq \sup_{x \in [0,1]} |W'(x)| \triangleq |W'| \tag{2.33}$$

Combining (2.31) - (2.33)

$$|b_{k+1}| \leq \begin{cases} |b_k|(1 + \theta|W'|) + L_1\theta^2 & \text{for } k\theta \leq N^* \\ |b_k|(1 - \frac{\theta\gamma}{2}) + L_2\theta^2 & \text{for } k\theta > N^* \end{cases} \tag{2.34}$$

Inequality (2.34) is of the same form as (2.24) and (2.26). Hence by the method of theorem 2.2 it immediately follows that

$$|b_k| = 0(\theta) \qquad\qquad \text{Q.E.D.}$$

As a consequence of Theorem 2.2 and 2.3 we have the following:

Corollary 2.3: If all the conditions of theorem 2.1 hold, then

$$\text{Var}[P(k) \mid P(0) = P] = 0(\theta), \text{ where}$$

var$(x|y)$ represents the variance of x given y.

Proof: From the definition

$$\text{Var}[P(k)] = E[(P(k) - E[P(k)])^2]$$

$$= E[(P(k) - f(k))^2]$$

$$- [E[P(k)] - f(k)]^2$$

From Theorem 2.2 and 2.3 it follows

$$\text{Var}[P(k)] \leq L_1\theta + L_2\theta^2 \leq L_3\theta$$

uniformly for all $P(0) = P \in [0,1]$ and $k \geq 0$. \qquad Q.E.D.

Remark 2.3

The case when $W'(P) < 0$ for all $P \in [0,1]$ deserves a special mention. In this case $W(P)$ can have at most one zero[4] in [0,1]. Now if $W(0) < 0$, then

$$E[P(1) \mid P(0) = 0] = P(0) + \theta W(0)$$

$$< 0$$

---

[4] For if there are two or more zeros, it would imply $W'(.)$ must change its sign which leads to a contradiction.

an impossibility. Thus $W(0) \geq 0$. Similarly $W(1) \leq 0$. Further if $W(0) = 0$ then $W(1) < 0$ and vice versa. Thus $W(p)$ has an unique zero, say $\beta \varepsilon [0,1]$ and $W'(\beta) < 0$ and this is exactly the additional condition that is needed in theorem 2.2. Further, the proof of the theorem 2.2 gets much simplified as indicated below:

$$\text{Define } R^* \underline{\Delta} \sup_{x \varepsilon [0,1]} W'(x) < 0 \tag{2.35}$$

$$h(\mu) \underline{\Delta} W(x + \mu(y-x))$$

From P2.4 it is clear that $h'(\mu)$ is continuous and from the fundamental theorem of calculus we have

$$W(x) - W(y) = h(1) - h(0) = \int_0^1 W'(x + \mu(y-x)) \, d\mu(x-y) \tag{2.36}$$

It follows from (2.14a), (2.35) and (2.36) that

$$R(x,y) \leq R* < 0 \tag{2.37}$$

Combining (2.37) and (2.17) - (2.19), the following inequality results

$$a_{k+1} \leq a_k[1 - 2|R^*|] + L_3\theta^2 \tag{2.38}$$

Setting $\theta < \dfrac{1}{2|R*|}$, iterating (2.38), as $a_0 = 0$, the conclusions of the theorem follows immediately.

Remark 2.4

　　a) When $\lambda(P) \equiv a$ and $\psi_j(P) \equiv ap_j$ and $0 < a < 1$ the algorithm is called the linear reward-penalty - L_{R-P}^E algorithm and for this choice

$$W(P) = a[c_2 - (c_1+c_2)P]$$

$$W'(P) = -a[c_1+c_2] < 0$$

　　b) if $\lambda(P) \equiv 0$ and $\psi_j(P) \equiv ap_j$, $0 < a < 1$ it results in linear inaction-[5] penalty L_{I-P}^E algorithm and

---

[5] Inaction refers to the fact that probabilities are not changed.

$$W(P) = a[(1-P)^2 c_2 - P^2 c_1]$$

$$W'(P) = -2a[(1-P) c_2 + P c_1] < 0$$

c)   When $\lambda(P) \equiv a$ and $\psi_j(P) = bp_j$ where $0 < a, b, < 1$.

$$W(P) = a P(1-P) (c_2-c_1) + b[(1-P)^2 c_2 - P^2 c_1]$$

Notice $W(0) = bc_2 > 0$ and $W(1) = -bc_1 < 0$ and as $W(P)$ is a quadratic in

P there exists an unique $\beta \in [0,1]$ and $W(\beta) = 0$, $W'(\beta) < 0$.   However $W'(P) \nless 0$

for all a and b, for all $P[0,1]$ and for all c_1 and c_2 as can be easily checked.

d)   Using the definition 1.3, it can be readily checked that all the

three special cases   referred to above belong to the class of distance

diminishing learning algorithms.

Remark 2.5:   As an immediate consequence of the above theorems we have

$$\lim_{k \to \infty} \sup \; | \; E[P(k)] - \beta_2 | = 0(\theta)$$

and

$$\lim_{k \to \infty} \sup \; Var[P(k)] = 0(\theta)$$

uniformly for all $P(0) = P \in [0,1]$

So far in our analysis we have confined our attention to the evolution

of the first and second moment of the process $\{P(k)\}$ for $k \geq 0$.   In the

following we set out to characterize the distribution of $P(k)$ after suitable

normalization.   To this end define

$$z(k) = \theta^{-\frac{1}{2}} [P(k) - f(k)] \tag{2.39}$$

Let $L[X]$ refer to the distribution of the random variable χ and $N[a,b]$

be the Normal distribution with mean a and variance b.   We have the following

Theorem 2.4:   If all the conditions of Theorem 2.1 are true then

(a) $L[z(k)] \sim N[0,g(t)]$

as $\theta \to 0$ and $k\theta \to t < \infty$ where $\sim$ indicates convergence in distribution and g(t) is the unique solution of the differential equation

$$g'(t) = 2 W'(f(t)) g(t) + s(f(t)) \qquad (2.40)$$

with $g(0) = 0$

(b) Further $\lim_{t \to \infty} L [z(k)] = N[0, g(\infty)]$

where $\theta \to 0$, $k\theta = t \to \infty$ and $g(\infty) = \dfrac{s(\beta_2)}{2|W'(\beta_2)|}$

The above theorem constitutes the central limit theorem for Markov

processes that evolve by small steps, and provides both transient (finite k)

and steady state ($k \to \infty$) characterization of the distribution of $z(k)$. Refer

to Table 2.3 to get an idea of the order of magnitude of $g(\infty)$.

Remark 2.6 Conditions (P2.4) and (P2.5) ensure that the solution to (2.40)

exists and is unique. However, it is well known [C5] that for the exis-

tence and uniqueness of the solution to (2.40) we only need that W'(P) and

s(P) satisfy the Lipschitz condition. In fact, no where in the proof of any

theorem given above do we need the differentiability of W'(P) and s(P).

These observations imply that (P2.4) and (P2.5) can be weakened to read as

follows:

(P'2.4) W(P) has bounded Lipschitz derivative in S_M.

(P'2.5) s(P) is Lipschitz in S_M.

The proof of the theorem is rather lengthy and will be given in various

steps. As a prelude we first establish the following two lemmas:

Lemma 2.2 Under the hypothesis of theorem 2.4, $E[z^2(k)]$ is bounded for all

$k \geq 0$ and $P(0) = P \epsilon[0,1]$.

Proof: Follows immediately from theorem 2.2.   Q.E.D.

Lemma 2.3: Under the hypothesis of the theorem 2.4, we have the following:

(a) $E[\delta z(k) \mid z(k)] = \theta W'(f(k)) \; z(k) + o\,(\theta)$

(b) $E[\,(\delta z(k)\,)^2 \mid 2(k)] = \theta s(f(k)\,) + o(\theta)$

(c) $E[\mid \delta z(k)\mid^3 \mid z(k)] = o(\theta)$

where $o(\theta)$ is such that $[E[o(\theta)]/\theta] \to 0$ as $\theta \to 0$ uniformly for all $P(0) = P$ and $k \le 0$.

Proof: From (P2.1) and (2.16) we have

$$E[\delta z(k) \mid z(k)] = \theta^{1/2}[E\{\,\delta P(k) \mid P(k)\} - \delta f(k)]$$
$$= \theta^{1/2}[W[P(k)] - W[f(k)]\,] + L\theta^{3/2} \qquad (2.41)$$

for some positive constant L.

As $W'(P)$ is Lipschitz we have

$$W(y) - W(x) - W'(x)\,(y-x) = \int_0^1 [W'[x + p(x - y)] - W'(x)]\,dp\,(y-x)$$
$$\le \frac{L_1}{2}\,(y-x)^2 \qquad (2.42)$$

From (2.42) and (2.41) it follows that

$$E[\delta z(k) \mid z(k)] = \theta W'[f(k)]z(k) + \theta^{3/2}0(\mid z(k)\mid^2) + 0(\theta^{3/2}) \qquad (2.43)$$

From lemma 2.2 as $z^2(k)$ is bounded, the conclusion (a) of this lemma follows.

To prove (b) we first observe

$$E[\,(\delta z\,(k)\,)^2 \mid z(k)\,] = \theta^{-1}\,\text{Var}\,[\delta P(k) \mid P(k)\,] + \{E[\delta z(k)\mid z(k)]\}^2 \quad (2.44)$$

From (2.43) we have

$$\{E[\,\delta z(k) \mid z(k)]\}^2 = o(\theta) \qquad (2.45)$$

From (P2.2)

$$\theta^{-1}\,\text{Var}\,[\delta P(k) \mid p(k)] = \theta s[P(k)]$$
$$= \theta s[f(k)] + \theta^{3/2}\,0(\mid z(k)\mid)$$
$$= \theta s[f(k)] + o\,(\theta) \qquad (2.46)$$

where the second equality is obtained by expanding $s[P(k)] \leq s[f(k)] +$

$|(P(k) - f(k))|L$ where L is the Lipschitz constant for $s(P)$. Combining

(2.44) - (2.46) we complete the proof of (b).

Finally, as $(a - b)^3 \leq 4[|a|^3 + |b|^3]$

$$E[|\delta z(k)|^3 \Big| z(k)] \leq 4\theta^{-3/2}[E[|\delta P(k)|^3 \Big| P(k)] + |\delta f(k)|^3]$$
$$\leq L_1 \theta^{3/2}$$

where the second inequality follows from (P2.3), (2.16) and the fact that $W(.)$

is bounded. This implies (c). (Q.E.D.)

Proof of Theorem 2.4:

Define $h(k,y) \underline{\Delta} E[\exp(iyz(k))]$

as the characteristic function of $z(k)$

Then

$$h(k+1,y) = E\left\{[\exp(iyz(k))] E[\exp(iy\delta z(k)) | z(k)]\right\} \qquad (2.47)$$

For any real x, using a third-order expansion

$$e^{ix} = 1 + ix - x^2/2 + O(|x|^3)$$

Thus

$$E[\exp(iy\delta z(k)) | z(k)]$$
$$= 1 + iy E[\delta z(k) \Big| z(k)] - y^2/2 E[(\delta z(k))^2 | z(k)]$$
$$+ |y|^3 0 E[|\delta z(k)|^3 | z(k)] \qquad (2.48)$$

Using the estimates of Lemma 2.3 in (2.48) and substituting the resulting

expression in (2.47), after simplification we get

$$\theta^{-1}[h(k+1, y) - h(k,y)] = W'[f(k)]y \frac{\delta h(k,y)}{\delta y} - s[f(k)]\frac{y^2}{2} h(k,y) \qquad (2.49)$$
$$+ \varepsilon_1(k,y)$$

where $\varepsilon_1(k,y) \to 0$ as $\theta \to 0$ for all k and y bounded and $h(0,y) = 1$

When θ is small, (2.49) suggests the following partial differential equation

$$\frac{\delta H(t,y)}{\delta t} = W'[f(t)]y \frac{\delta H(t,y)}{\delta y} - s(f(t)) \frac{y^2}{2} H(t,y) \qquad (2.50)$$

This is a linear partial differential equation of first order and can be
solved explicitly using the method of characteristics [C3] [C5]. In order to
keep the length of the already lengthy proof within reasonable limits, at
this point without loss of continuity we only indicate the major steps here.
However, proof technique is interesting in itself. The details are given in
appendix to this chapter.

It can be shown (see appendix) that

(i)   The solution of (2.50) for which $H(0,y) = 1$ is the characteristic
      function

$$H(t,y) = \exp[-g(t) \frac{y^2}{2}] \qquad (2.51)$$

of the normal distribution with mean zero and variance $g(t)$.

(ii) $$\lim_{\substack{k\theta \to t \\ \theta \to 0}} h(k,y) = H(t,y) \qquad (2.51a)$$

Now substituting (2.51) in (2.50) we get

$$g'(t) = 2 W'(f(t)g(t) + s(f(t)).$$

This completes the proof of (a) of Theorem 2.4.

To prove the part (b) let us first recall from (2.14a)-(2.14c) that

$$|f(t) - \beta_2| \le e^{-\gamma t}$$

As $W'(.)$ and $s(.)$ are Lipschitz, we have

$$|W'(f,k)) - W'(\beta_2)| \le L| f(k) - \beta_2| < L e^{-\gamma k\theta} \qquad (2.52)$$

and similarly

$$|s(f(k)) - s(\beta_2)| < L e^{-\gamma k\theta} \qquad (2.53)$$

where θ is small. In view of (2.52) and (2.53) for large values of k and
θ small (2.49) can be rewritten as follows

$$\theta^{-1}[h(k+1,y) - h(k,y)] = W'(\beta_2) \; y \; \frac{\partial h(k,y)}{\partial y} - s(\beta_2) \; \frac{y_2}{2} \; h(k,y)$$

$$+ \; \varepsilon_1(k,y) + \varepsilon_2(k,y) \qquad (2.54)$$

where $\varepsilon_2(k,y)$ reflect the error in replacing $W'(f(k))$ and $s(f(k))$ with $W'(\beta_2)$ and $s(\beta_2)$ $\varepsilon_2(k,y) \to 0$ as $k\theta \to \infty$, $\theta \to 0$ and y bounded.

Define $\qquad v(y) \triangleq \exp \left(\frac{y^2 \sigma^2}{2} \right)$

$$\left. \begin{array}{c} \sigma^2 \triangleq \dfrac{s(\beta_2)}{2 \; W'(\beta_2)} \\[2em] u(k,y) \triangleq v(y) \; h(k,y) \\[1em] \text{We have the identity } v(y) \dfrac{\partial h(k,y)}{\partial y} = \dfrac{\partial u(k,y)}{\partial y} - \sigma^2 y \; u(k,y) \end{array} \right\} \qquad (2.55)$$

Multiply (2.54) by $v(y)$ and on simplification using (2.55) we get

$$\theta^{-1}[u(k+1,y)] = W'(\beta_2)y \; \frac{\partial u(k,y)}{\partial y} + E_1(k,y) + E_2(k,y) \qquad (2.56)$$

where $E_i(k,y) \to 0$ as $k\theta \to \infty$, $\theta \to 0$ and y bounded.

Now (2.56) suggests the partial-differential equation

$$\frac{\partial U(t,y)}{\partial t} = W'(\beta_2) \; y \; \frac{\partial U(t,y)}{\partial y} \qquad (2.57)$$

It is well known from the theory of partial differential equations that for

any constant g, $(t, \; g \; e^{-W'(\beta_2) \; t})$ is a characteristic base curve of the

equation (2.57) and the solution $U(.,.)$ is a constant along any characteristic

curve, that is

$$\frac{d}{dt} \; U(t, \; g \; e^{-W'(\beta_2)t}) = 0$$

Define $y_j = (1 + \theta a)^j \; y$ where $a \triangleq W'(\beta_2)$

Choosing $0 < \theta < \frac{1}{|a|}$, we have for finite y

$$y_j \leq e^{a\theta j} |y|$$

For any $0 \leq n \leq N$, define

$$u_n \triangleq u(N, y_{N-n})$$

It is shown in Appendix that $u_N - u_\ell \to 0$ as $\Theta \to 0$ and $\ell\Theta \to \infty$ and $u_\ell \to 1$ as

$\theta(N-\ell) \to \infty$. Now choosing $\ell = [M/2]$ (the integer part of $M/2$) we see that

$u_N \to 1$, and $h(N,y) \to \exp[-\frac{y^2\sigma^2}{2}]$, the characteristic function of $N(0,\sigma^2)$ as

$\theta \to 0$ and $N\theta \to \infty$.   Q.E.D.

It follows from the above analysis that the value of β_2 (the zero of

the function $W(P)$) and the behavior of $W(P), s(P)$ in the neighborhood of β_2

very much control the asymptotic properties of $z(k)$ and hence of $p(k)$.

Our next and final theorem of this section establishes that there exists

proper choice of parameters such that N_{R-P}^E algorithm given by (2.1a) is

ε-optimal.

Theorem 2.5: Let

(1) $\lambda(.)$ and $\alpha\psi_j(.)$ satisfy the conditions (C2.1)-(C2.3) where α is a
     real and positive and

(2) $(1 - P)\ \psi_2\ [P] \gtrless P\ \psi_1 [P]$ for $P \lessgtr 1/2$

     Then for every $\varepsilon > 0$, there exists an α^* such that for all $\alpha \leq \alpha^*$,

     the algorithm (2.1a) is ε-optimal.

Proof:

Step 1: Under the conditions of this theorem, it follows from theorem 2.1,

         Corollary 2.1 and 2.2 that $W(\alpha,P)$ (as defined in 2.9) has an unique

         zero $\beta(\alpha)\ \varepsilon\ (0,1)$ such that $W'[\alpha,\beta(a)] < 0$ and $\beta(\alpha) \to 1$ as $\alpha \to 0$.

Step 2: It follows from Remark 2.5 that

$$\lim_{k \to \infty} \sup \ | \ E[P(k)] - \beta(\alpha) \ | = 0(\theta)$$

Step 3: From Steps 1 and 2, for any given $\delta > 0$ there exists $0 < \alpha^* < 1$ and

         $\theta^* < 1$ such that for all $\alpha < \alpha^*$ and $\theta < \theta^*$

$$\lim_{k \to \infty} \sup \left| E[P(k)] - 1 \right| < \delta$$

Step 4: As

$$\eta(k) = d_1 + (d_2 - d_1) E[1 - P(k)]$$

we have

$$\lim_{k \to \infty} \sup \left| \eta(k) - d_1 \right| < \left| d_2 - d_1 \right| \delta$$

Since $d_1 \neq d_2$, choose $\delta = \dfrac{\varepsilon}{|d_2 - d_1|}$ and the theorem follows from definition (1.1).

Q.E.D.

Remark 2.7: (a) If $\lambda[.] \equiv 0$, then $W(P) \equiv W^P(P)$ and all theorems and corollaries still remain true. In this case as $W(\alpha, P) = \alpha W(P)$, the zero of $W(P)$ cannot be placed anywhere we like but fixed uniquely by the choice of $\psi_j(.)$. However, if $\psi_j(.)$ satisfies the condition (2) of the theorem 2.5, it can be shown (see problem 2.2) that the algorithm (2.1a) is still "expedient".

(b) If $\psi_j[P] \equiv 0$ for all j and $\lambda[P]$ satisfies conditions C2.1 - C2.3, the properties of the algorithm (2.1a) undergo a dramatic change and are called "absolutely expedient algorithms". These algorithms are the topic of Chapter 3.

2.4 An Alternate Characterization of z(k):

Recall that

$$z(k) = \theta^{-\frac{1}{2}} [P(k) - f(k)], \; k = 0,1,2,\ldots$$

where $f(k) \triangleq f(k\theta)$ and $f(t)$ is defined by (2.13). From lemma 2.3 we know that the Markov process $\{z(k)\} \; k \geq 0$ satisfies the following conditions:

$$\lim_{\theta \to 0} \; \frac{1}{\theta} \; E[\; (\delta z(k) \;)^3 \; | z(k) \;] = 0$$

$$\lim_{\theta \to 0} \; \frac{1}{\theta} \; E \; [\; \delta z(k) \; | \; z(k) \;] = \; W'(f(k)) \; z(k)$$

$$\lim_{\theta \to 0} \; \frac{1}{\theta} \; E \; [\; (\delta z(k) \;)^2 \; | \; z(k) \;] = \; s[f(k)]$$

It is well known that any Markov process in discrete time satisfying the above conditions can be well approximated by Markov diffusion process [G4 page 64-65]. To this end, define the following piece-wise linear inter-polation of z(k):

$$\hat{z}_\theta(t) = \frac{(t-k\theta)}{\theta} \; z(k+1) + \frac{((k+1)\theta-t)}{\theta} \; z(k) \tag{2.58}$$

for $t \in [k \; \theta,(k+1)\theta]$, $k = 0,1,2,\ldots$ Notice $\hat{z}_\theta(t) \; \epsilon \; C \; [0, \infty)$, the space of continuous functions on $[0, \infty)$. Hence $\hat{z}_\theta(t)$ as $\theta \to 0$ and $k \theta \to t$ converges in distribution to the Markov diffusion process $\hat{z}(t)$ that evolves according to the following stochastic differential equation

$$d \; \hat{z}(t) = W'(f(k)) \; \; \hat{z}(t) \; dt + b[f(t)] \; dB(t) \tag{2.59}$$

where $b^2 \; [f(t)] = s[f(t)]$ and $B(t)$ is the standard Brownian Motion process [G3] [G4] [A2].

The above differential equation is a linear stochastic differential and hence we can solve (2.59) explicitly. From theorem (8.2.2) in Arnold [A2 pp 129-130] it can be verified that

$$\hat{z}(t) = \Phi(t) \int_0^t \Phi(s) \, b[f(t)] \, d \, B(t)$$

where

$$\Phi(t) = \exp \left[\int_0^t \frac{dW[f(u)]}{du} \, du \right]$$

2.5 Simulations (M=2)

In this section we present a number of simulation experiments which illustrate the theory outlined in the above section. In each of the examples presented below $E[P(k)]$ is calculated as the numerical average of fifty sample runs.

Example 2.2: $\lambda(P) = 0.1$

$$\psi_1(P) = \alpha P \, [.6 - .5P]$$

$$\psi_1(P) = \alpha(1-P) \, [.6 - .5(1-P)]$$

If $\alpha = 0.01$, $d_1 = 0.6$; $d_2 = 0.35$ it can be shown that $\beta(\alpha) = 0.983$

$W' \, [\alpha, \beta(\alpha)] = -0.0232$ and

$\frac{s[\beta(\alpha)]}{2|W'(\alpha, \beta(\alpha))|} = .1287 \times 10^{-2}$. The evolution $E[P(k)]$ for different

values of the step length parameter θ is given in Table 2.4.

Example 2.3: This is same as example 2.2 except that $\alpha = 1.0$. In this case it can be seen that $\beta(1) = 0.618$, $W'[1, \beta(1)] = -0.2290$ and

$\frac{s[\beta(1)]}{2|W'[1, \beta(1)]|} = .3297 \times 10^{-1}$. Refer Table 2.5 for values of $E[P(k)]$.

Table 2.4

k	E[P(k)]	
	θ = 0.50	θ = 0.25
0	0.5000	0.5000
100	0.7460	0.6441
500	0.9836	0.9399
1000	0.9867	0.9775
1500	0.9873	0.9841
2000	0.9880	0.9865
2500	0.9809	0.9836
3000	0.9837	0.9823

Table 2.5

k	E[P(k)]			
	θ = 0.50	θ = 0.25	θ = 0.10	θ = 0.05
0	0.5000	0.5000	0.5000	0.5000
100	0.5853	0.5819	0.5820	0.5657
500	0.6200	0.6150	0.6156	0.6189
1000	0.6611	9.6463	0.6270	0.6203
1500	0.6624	0.6314	0.6202	0.6191
2000	0.5717	0.5931	0.6055	0.6092
2500	0.6217	0.5956	0.6289	0.6232
3000	0.5780	0.5956	0.6157	0.6173

Example 2.4: $\lambda[P] = 0.1$

$\psi_1 [P] = \alpha P$

$\psi_2 [P] = \alpha (1-P)$

$d_1 = 0.60$, $d_2 = 0.35$, $\alpha = 0.01$; $\beta(\alpha) = 0.869$; $W'[\alpha,\beta(\alpha)] = -0.02710$.

Refer Table 2.6 for $E[P(k)]$.

Table 2.6

k	E[P(k)]	
	$\theta = 0.10$	$\theta = 0.20$
0	0.5000	0.5000
100	0.5615	0.6166
500	0.7411	0.8309
1000	0.8280	0.8608
1500	0.8570	0.8623
2000	0.8659	0.8697
2500	0.8610	0.8555
3000	0.8641	0.8616

Examples 2.2 and 2.3 clearly illustrate the role of the parameter α, namely by decreasing its value, (everything else remaining the same) we can improve the performance of the learning algorithm from "expediency" to ε-optimality, provided $\lambda[P]$ is not identically zero.

2.6 Analysis and Simulation: General Case $M \geq 2$

The analysis of the Ergodic algorithm for the general case of $M(\geq 2)$ coins essentially follow the same lines of those presented in the above sections except that we have to establish the multidimensional analogs of the lemmas and theorems stated above. We shall only state the theorems and leave the verification as exercises.

The first step involves finding conditions under which $\underline{W}(p)$ (refer P2.1) will have a unique zero in S_M. Let

(C2.4) $P_1 \psi_1 (p) \gtrless P_2 \psi_2 (p) \gtrless P_3 \psi_3 (p) \gtrless \cdots \cdots \gtrless P_M \psi_M (p)$

when

$$P_1 \gtrless P_2 \gtrless P_3 \gtrless \cdots \cdots \gtrless P_M$$

Theorem 2.6: Assume (C2.1) - (C2.4). Then

(a) There exists a unique $\hat{p} \epsilon S_M$ such that

$$\underline{W}^P (\hat{p}) = 0$$

and

(1) $\hat{p} = (\frac{1}{M} , \frac{1}{M} , \ldots \frac{1}{M})^T$ if $d_1 = d_2 = \ldots = d_M$

(2) $\hat{p}_i > \hat{p}_j$ for all $j \neq i$ if $d_i > d_j$ for all $j \neq i$

(b) $\lambda(p) \neq 0$ implies there exists a unique p^* such that

$$\underline{W} (p^*) = 0$$

and

(1) $p^* = (\frac{1}{M}, \frac{1}{M}, \ldots \frac{1}{M})^T$ if $d_1 = d_2 = \ldots = d_M$

(2) $p_i^* > \hat{p}_j > \frac{1}{M}$ where $d_i > d_j$ for all $j \neq i$.

and

(3) There exists proper choice of $\underline{\psi}_i (p)$ i = 1 to M such that p_i^* can be made as close to unity as desired.

49

Proof: Recall that

$$\underline{W}(p) = (W_1(p), W_2(p) \quad \cdots \cdots \cdots W_M(p))^T \text{ where}$$

$$W_i(p) = W_i^R(p) + W_i^P(p)$$

$$W_i^R(p) = \lambda(p) \, p_i \sum_{j \neq 1} p_j (d_i - d_j)$$

and

$$W_i^P(p) = \frac{1}{(M-1)} \sum_{j \neq 1} [p_j \psi_j(p) c_j - p_i \psi_i(p) c_i]$$

We shall record some obvious properties of these functions:

(A) $\underline{W}^R(p) = 0$ if $p \in V_M$

(B) $\underline{W}^P(p) \neq 0$ for all $p \in V_M$

(C) $\underline{W}(p) \neq 0$ for all $p \in V_M$

(D) $W_i(e_j) \begin{cases} > 0 & \text{for all } i \neq j \\ < 0 & \text{for } i = j \end{cases} \quad i = 1, 2, \ldots, M$

where e_i is the ith unit vector.

If $d_1 = d_2 = \ldots = d_M = d = 1 - c$ (Say) then it is easily verified that

$$W_i(p) = W_i^P(p) = \frac{c}{m-1} \sum_{j \neq 1} [p_j \psi_j(p) - p_i \psi_j(p)]$$

From this and (C2.4), the conclusions (a.1) and (b.1) of this theorem immediately follow.

For concreteness let $d_1 > d_2 > \ldots > d_M$ (refer assumption A1.1). Condition (C2.4) and the properties of $W_i^P(p)$ listed above imply that there exists a unique $\hat{p}$ such that $W_i^P(p) = 0$ and $\hat{p}_1 > \hat{p}_j$ for all $j \neq 1$. In other words, the effect of the assumption 1.1 is to translate the zero of $\underline{W}^P(p)$ towards the vertex corresponding to the unit vector e_1 of the simplex S_M. From this (a.2) follows.

We shall only indicate the idea of the proof of b.2 and b.3. Under assumption 1.1, if $\lambda(p) \not\equiv 0$, the effect of adding $W_i^R(p)$ to $W_i^P(p)$ is to further shift or translate the zero p^* of $\underline{W}(p)$ such that $p_1^* > \hat{p}_1$. Further if we replace $\psi_i(p)$ by $\alpha\psi_i(p)$ where $0 < \alpha < 1$, then the zero $p^*(\alpha)$ of

$$\underline{W}(\alpha, p) = (W_1(\alpha, p), w_2(\alpha, p) \ldots W_M(\alpha, p))^T$$ where $W_i(\alpha, p) = W_i^R(p) + \alpha W_i^P(p)$

is such that $p_1^*(\alpha) \to 1$ as $\alpha \to 0$ provided assumption 1.1 holds and $\lambda \not\equiv 0$.

Q.E.D.

The following examples illustrate the above theorem

<u>Example 2.5</u>: Let $D = (.8, .4, .15)$; $\lambda(p) = 0.1$ and $\psi_i(p) = B_1 p_i$ where $0 < B_1 < 1$. For values of $B_1 = 0.1$ and 0.01, the plots of $W_i(p)$ for $i = 1,2,3$ are given in the figures 2.1 and 2.2 from which we have the following:

B_1	P_1^*	P_2^*	P_3^*
0.10	.7405	.1465	0.1130
0.01	.960	.025	0.015

<u>Example 2.6</u>: Let $D = (.8, .4, .15)$ $\lambda(p) = 0.1$ and $\psi_i(p) = B_1 p_i [B_2 - B_3 p_i]$. See figure 2.3 for plots of $W_i(p)$ $i = 2,3$ where $B_2 = 0.6$ and $B_3 = 0.5$.

B_1^*	P_1^*	P_2^*	P_3^*
.05	.980	.012	.008

As $\sum_{i=1}^{M} P_i = 1$ and $\sum_{i=1}^{M} w_i(p) = 0$ for all $p \varepsilon S_M$ in the following we will only consider the (M-1) independent components of the vectors p and $\underline{W}(p)$. Without loss of generality we consider the first (M-1) components. Define

$$\tilde{p} = (p_1, p_2, \ldots, p_{M-1})^T$$

$$\underline{\tilde{W}}(\tilde{p}) = (W_1(\tilde{p}), W_2(\tilde{p}), \ldots, W_{M-1}(\tilde{p}))^T$$

Notice $\{\tilde{p}(k)\}$ as defined above is a Markov process if $\{\tilde{p}(k)\}$ is. Let

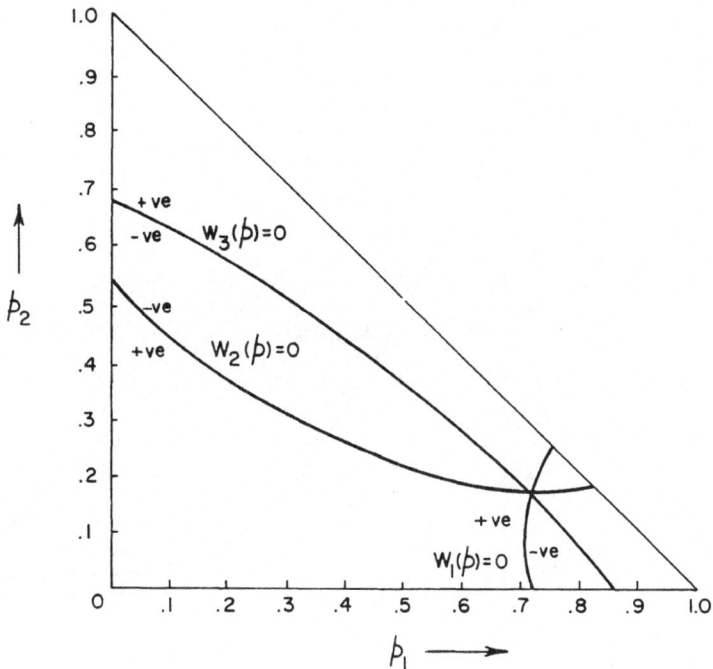

Figure 2.1

P* = (.7405, .1465, .1130)

D = (.8, .4, .15)

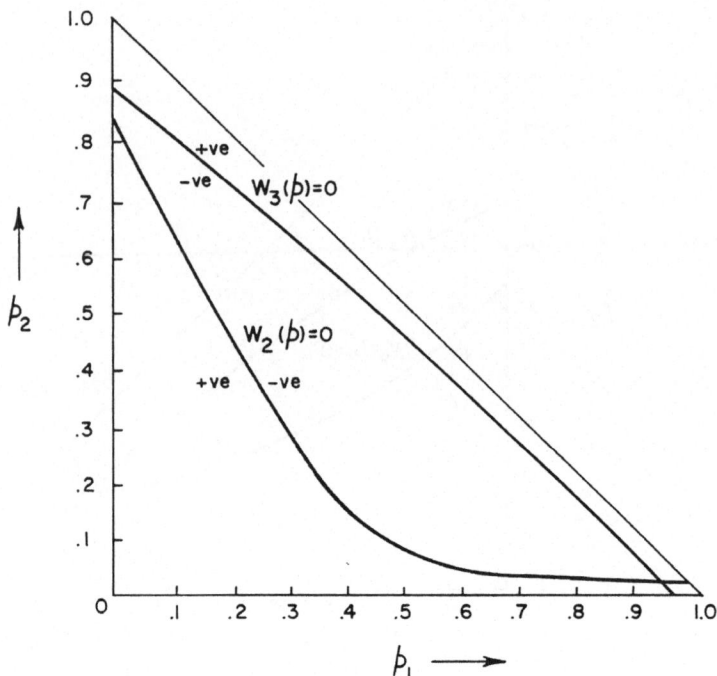

Figure 2.2

$$P^* = (.980, .012, .008)$$
$$D = (.8, .4, .15)$$

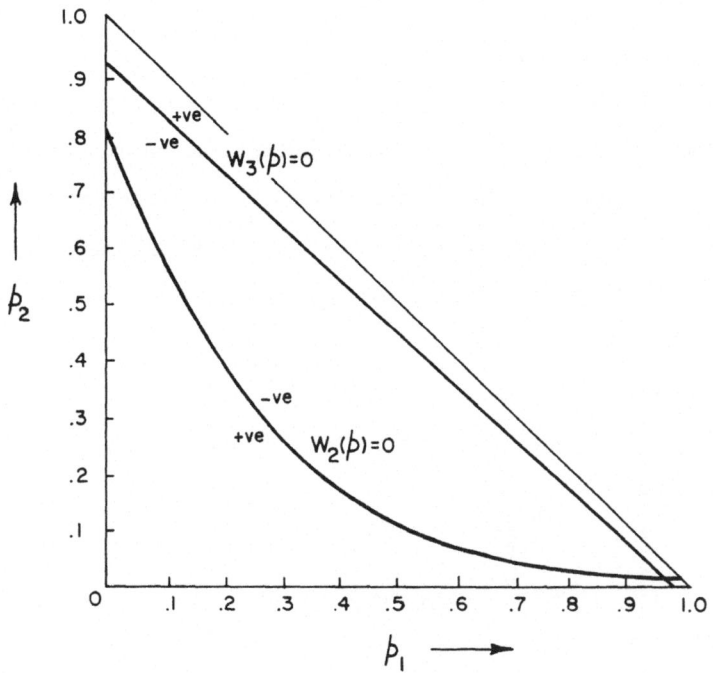

Figure 2.3

P* = (.960, .025, .015)

D = (.8, .4, .15)

$$\tilde{S}_M = \{\tilde{p} \mid \tilde{p} = (p_1, p_2, \ldots, p_{M-1})^T, \ p_M = 1 - \sum_{i=1}^{M-1} p_i \ \\ \text{and } p = (p_1, p_2, \ldots, p_M) \ \epsilon \ S_M \} \quad (2.61)$$

and let $\tilde{S}_M^o$ be the interior of $\tilde{S}_M$.

$$\mu_k(\tilde{p}) = E[\ \tilde{p}(k) \mid \tilde{p}(o) = \tilde{p}] \ \\ V_k(\tilde{p}) = E[\ \mid \tilde{p}(k) - \mu_k(\tilde{p})\mid^2 \ \mid \tilde{p}(o) = p] \quad (2.62)$$

where $|.|$ is the Euclidean norm.

The following theorem summarizes the properties of the Markov process $\{\tilde{p}(k)\}$ and is a combination (of the multidimensional analogues) of many theorems we have established in the previous sections for the case of $M = 2$.

Theorem 2.7: Under the assumptions (C2.1) - (C2.4), the following are true:

   I.   There exists a unique $\tilde{p}^* \ \epsilon \ \tilde{S}_M^o$ such that [6]

        a) $\underline{\tilde{W}}(\tilde{p}^*) = 0$

        b) $[\tilde{p} - \tilde{p}^*, \ \underline{\tilde{W}}(\tilde{p})\ ] < 0$ $\quad (2.63)$

        for all $\tilde{p} \neq \tilde{p}^*$ and $\tilde{p} \epsilon \tilde{S}_M$

        and

        c) $[Z, \ \dfrac{d\tilde{W}(\tilde{p}^*)}{d\tilde{p}} \ Z\ ] < 0$ $\quad (2.64)$

        for all $Z \ \epsilon \ R^{M-1}$, the Euclidean (M-1) dimensional space and $Z \neq 0$.

   II. $V_k(\tilde{p}) = 0(\theta)$ $\quad (2.65)$

        uniformly for all $\tilde{p} \ \tilde{S}_M$ and $k \geq 0$

   III. For any $\tilde{p} \ \tilde{S}_M$, the differential equation

        a) $\dfrac{d\tilde{f}}{dt} = \tilde{W}[\ \tilde{f}(t)\ ]$ $\quad (2.66)$

        has a unique solution $\tilde{f}(t)$ with $\tilde{f}(0) = \tilde{p}$ and

---

[6] $[x,y]$ refers to the usual inner product of the vectors z and y.

b) $| \tilde{f}(k\theta) - \mu_k(\tilde{p})| = 0(\theta)$ (2.67)

<u>uniformly for all</u> $\tilde{p} \in \tilde{S}_M$ <u>and</u> $k \geq 0$

IV. a) $L(\tilde{Z}_k) \sim N[0, \tilde{g}(t)]$

<u>as</u> $\theta \to 0$ <u>and</u> $k\theta \to t < \infty$ <u>where $\sim$ indicates convergence in</u>

<u>distribution</u> and $g(t)$ <u>is the unique solution of</u>

<u>and</u> $\dfrac{dg(t)}{dt} = \dfrac{d\tilde{W}(t)}{dt} g(t) + g(t)(\dfrac{dW(t)}{dt})^T + \tilde{s}[\tilde{f}(t)]$ (2.68)

$\tilde{Z}_k = \theta^{-\frac{1}{2}} [\tilde{p}(k) - f(\tilde{k}\theta)]$ (2.69)

<u>and</u>

b) $\lim\limits_{k \to \infty} L(\tilde{Z}_k) = N[0, \tilde{g}(\infty)]$

<u>where the matrix</u> $\tilde{g}(\infty)$ <u>is obtained as the unique solution of the</u>

<u>linear system</u>

$\dfrac{d\tilde{W}(\tilde{p}^*)}{d\tilde{p}} \tilde{g}(\infty) + \tilde{g}(\infty) \dfrac{d\tilde{W}(\tilde{p}^*)}{d\tilde{p}} T + \tilde{s}(\tilde{p}^*) = 0$ (2.70)

<u>where $\tilde{s}(\tilde{p})$ is obtained by deleting the</u> $(M-1)^{th}$ <u>row and column of</u>

$s(p)$ [refer P2.2].

Theorem 2.7 stated above is a combination of many theorems from Norman [N10] [N13] (2.63) follows from theorem 2.6 but the proof of (2.64) is non-trivial. From (2.63) and (2.64) it follows that [L11]   the differential equation (2.66) is uniformly asymptotically stable and hence $\lim\limits_{t \to \infty} \tilde{f}(t) = \tilde{p}^*$. Using this fact and (2.67) and by proceeding along the lines of theorem (2.5) we can indeed show that <u>ε-optimality</u> follows. Further (2.70) gives the asymptotic covariance matrix of $\tilde{Z}(k)$ and hence of $\tilde{p}(k)$.

In the following we shall illustrate the above theorem through a variety of examples. In all the examples $E[p_1(n)]$ is calculated as the numerical average over fifty sample runs.

Example 2.7: This is the same as example 2.5 with B1 = 0.01.

It is seen from example 2.5 that $p^* = (.96, .025, .015)$.

The matrix $\dfrac{d\tilde{W}(p^*)}{d\tilde{p}}$ is given below and the results of simulations are given in table 2.7.

$$\frac{d\tilde{W}(p^*)}{d\tilde{p}} = \begin{bmatrix} -0.0644 & -0.0240 \\ .00023 & -0.0391 \end{bmatrix}$$

The trace of this matrix T = -0.1035 and determinant Δ = 0.002512. Hence $T^2 - 4\Delta > 0$ and both the eigenvalues of this matrix are real and negative [B6, Chapter 3].

Table 2.7

$\lambda(p) = 0.1, \ \psi_i(p) = 0.01 \ p_i$

k	$E[p_1(k)]$	
	$\theta = 0.25$	$\theta = 0.10$
0	0.3333	0.3333
100	0.6352	0.4586
200	0.8258	0.5778
300	0.8943	0.6813
400	0.9313	0.7579
500	0.9688	0.8176
1000	0.9566	0.9397
1500	0.9556	0.9529
2000	0.9597	0.9553
2500	0.9633	0.9597
3000	0.9617	0.9620

When $B_1 = 0.10$ from example 2.5, $p^* = (.7405, .1465, .1130)$ and values of $E[p_1(k)]$ are given in the following table 2.8. In this case

$$\frac{d\ \tilde{W}(p^*)}{d\ \tilde{p}} = \begin{bmatrix} -0.09188 & -.01933 \\ -0.01465 & -0.05992 \end{bmatrix}$$

Trace T = -0.1518, determinant Δ = 0.00522 and $T^2-4\Delta>0$ and hence the eigenvalues are real and negative.

Table 2.8

$\lambda(p) = 0.1, \ \psi_i(p) = 0.1\ p_i$

k	$E[p_1(k)]$	
	$\theta = 0.25$	$\theta = 0.10$
0	0.3333	0.3333
100	0.6239	0.4834
200	0.7143	0.5836
300	0.7333	0.6497
400	0.7289	0.6861
500	0.7220	0.7095
1000	0.7233	0.7278
1500	0.7136	0.7229
2000	0.7599	0.7455
2500	0.7268	0.7291
3000	0.7154	0.7285

Example 2.8: This is the same as example 2.6. See table 2.9. Recall that $p^* = (.980, 0.012, 0.008)$.

Table 2.9

$\lambda(p) = 0.1$, $\psi_i(p) = B_1 p_i [B_2 - B_3 p_i]$, $B_1 = 0.05$, $B_2 = 0.6$, $B_3 = 0.5$

n	$E[p_1(k)]$	
	$\theta = 0.25$	$\theta = 0.10$
0	0.3333	0.3333
100	0.6400	0.4632
200	0.8341	0.5840
300	0.9060	0.6858
400	0.9477	0.7611
500	0.9666	0.8204
1000	0.9717	0.9512
1500	0.9783	0.9691
2000	0.9779	0.9764
2500	0.9748	0.9771
3000	0.9787	0.9796

2.7   Comments and Historical Remarks:

Much of the material of this chapter follows Norman [N10]-[N16] and is
now well known as the Theory of Markov process that evolve by small steps.
Our presentation in particular follows [N13], [N15] and L [7]. In [N16]
Norman has extended these results to a general class of discrete time
stochastic processes.

The idea of approximating the evolution of the mean of a Markov process
by the solution of an ordinary differential equation has been well exploited
in the literature [K6] [K7].   Kurtz [K6] [K7] considers the problem of
approximating the limiting behavior of trajectories of certain class of
jump Markov processes. See also Meerkov [M5]. Recently in the context of
adaptive control in systems theory Kushner [K12] and Ljung [L12] have
independently shown that the trajectories of a class of recursive stochastic
algorithms converge with probability one to the asymptotically stable
solutions of certain associated ordinary differential equation. We will
exploit this idea especially the method due to Kushner [K11]   in Chapter 4
for the analysis of time varying learning algorithms and more will be said
about these methods in that chapter.

The method of proof of theorem 2.4 is an adaptation of those due to
Rosen [R5] who develops a central limit theorem for sums of dependent
random variables.

2.8 APPENDIX

A. Proof of (2.51) and (2.51a):

Define B(t) as the solution of the ordinary differential equation

$$\frac{dB(t)}{dt} = - W' [f(t)] B(t) \tag{2.A1}$$

where B(0) = 1. That is,

$$B(t) = \exp [- \int_0^t W' [f(u)] du] \tag{2.A2}$$

Therefore, from (2.50) we get, for any real ξ

$$\frac{dH[t,\xi B(t)]}{dt} = - s[f(t)] \frac{[\xi B(t)]^2}{2} H[t, \xi B(t)] \tag{2.A3}$$

Since $H(0,\xi) = 1$, we obtain the solution of (2.A3) as

$$H[t, \xi B(t) = \exp [- \frac{1}{2} \int_0^t [\xi B(u)]^2 s [f(u)] du] \tag{2.A4}$$

for all t and ξ. Now choosing $\xi = yb^{-1}(t)$ we obtain (2.51).

To prove (2.51a) let $\xi_k \triangleq \xi B(k\theta)$ and $H_k \triangleq H(k\theta,\xi_k)$, $h_k \triangleq h(k,\xi_k)$ and $f(k) = f(k\theta)$. As B(t) is bounded for $t < \infty$, we have ξ_k bounded for all k such that $\theta \to 0$ and $k\theta \to t$.

Taylor expansion of H_{k+1} gives

$$H_{k+1} = [H[(k+1)\theta, \xi B[(k+1)\theta]]$$

$$= H[k\theta, \xi B(k\theta)] + \theta \frac{dH[k\theta,\xi B(k\theta)]}{dt} + 0 (\theta^2) \tag{2.A5}$$

Combining (2.A5) and (2.A3) we get

$$H_{k+1} - H_k = - \frac{1}{2} \theta s[f(k)] \xi_k^2 H_k + 0 (\theta^2) \tag{2.A6}$$

Also

$$h_{k+1} - h_k = h[k+1, \xi_{k+1}] - h[k, \xi_k]$$

$$= [h(k+1,\xi_{k+1}) - h(k+1, \xi_k)] + [h(k+1,\xi_k) - h(k,\xi_k)] \tag{2.A7}$$

But using Taylor expansion and (2.A3) we obtain

$$h(k+1,\xi_{k+1}) - h(k+1,\xi_k) = \theta \, \frac{\partial h(k+1)\,\xi_k)}{\partial y} \, \{- \xi_k \, W'[f(k)]\} \tag{2.A8}$$

and from (2.49)

$$h(k+1,\xi_k) - h(k,\xi_k) = \theta W'[f(k)]\xi_k \, \frac{\partial h[k,\xi_k]}{\partial y} - \frac{1}{2}\theta \, s[f(k)]\xi_k^2 \, h(k,\xi_k)$$
$$+ \varepsilon_1 (k,\xi_k) \tag{2.A9}$$

where $\varepsilon_1(k,\xi_k) \to 0$ as $\theta \to 0$.

Substituting (2.A8) and (2.A9) into (2.A7) we obtain

$$h_{k+1} - h_k = -\frac{1}{2} \theta s \, [f(k)]\xi_k^2 \, h_k + 0(\theta) \tag{2.A10}$$

Subtracting (2.A10) from (2.A6)

$$H_{k+1} - h_{k+1} = [1 - \frac{1}{2}\theta \, s[f(k)]\xi_k^2] \, (H_k - h_k) + 0(\theta) \tag{2.A11}$$

Notice $s[f(k)]\xi_k^2$ is non-negative and bounded. (2.A11) implies that $(H_k-h_k) \to 0$

as $\theta \to 0$. But $H_k \to H[t,\xi B(t)]$ and $h_k - h(k,\xi B(t) \to 0$ as $k\,\theta \to t$, hence

$h(k,\xi B(t) \to H[t,\xi B(t)]$ as $k\theta \to t$, $k \to \infty$ and $\theta \to 0$ for all ξ and $t < \infty$.

By changing the variable $\xi = yB^{-1}(t)$ and by continuity theorem we

complete the proof of (2.51a).

B. Proof of $u_N - u_\ell \to 0$ as $\theta \to 0$ and $\ell\theta \to \infty$:

Define $\Delta u_n = u(n,y_{N-n}) - u(n-1, y_{N+1-n})$ for $n \geq 1$

$$\overset{\Delta}{=} A_n - B_n$$

where

$$A_n = [u(n,y_{N-n}) - u(n-1, y_{N-n})]$$

$$B_n = [u(n-1, y_{N+1-n}) - u(n-1,y_{N-n})]$$

Thus

$$|u(N,y) - u(N,Y_{N-K}) | \leq \sum_{n=k+1}^{N} |A_n - B_n |$$

From (2.56) we have

$$A_n = \theta a y_{N-m} \frac{\partial u[n-1, y_{N-n}]}{\partial y} + E_1 [n, y_{N-n}] + E_2 [n, y_{N-n}] \qquad (2.A11)$$

By Taylor-expansion

$$B_n = u(n-1, y_{N+1-m}) - u(n-1, y_{N-n})$$

$$= (y_{N+1-n} - y_{N-n}) \frac{\partial u[n-1, y_{N-n}]}{\partial y} + F_{n-1}$$

$$= \theta a y_{N-n} \frac{\partial u[n-1, y_{N-n}]}{\partial y} + F_{n-1} \qquad (2.A12)$$

where, for some positive constant K and from lemma 2.2

$$F_{n-1} \leq \frac{1}{2} \left| y_{N+1-n} - y_{N-n} \right|^2 \max_{|x| \leq |y|} \left| \frac{\partial^2 u[n-1, x]}{\partial y^2} \right|$$

$$\leq k \theta^2 e^{2a\theta(N-n)} \qquad (2.A13)$$

Subtracting (2.A12) from (2.A11) we obtain

$$A_n - B_n = E_1 [n, y_{N-n}] + E_2 [n, y_{N-n}] - F_{n-1} \qquad (2.A14)$$

But $E_1(.,.)$, $E_2(.,.)$ and $F_{n-1}(.)$ all tend to zero as $\theta \to 0$, $n\theta \to \infty$, it follows that $u(N,y) - u(N,y_{N-k}) \to 0$ as $\theta \to 0$ and $k\theta \to \infty$.

Finally, using lemma 2.2, we obtain

$$\left| h(k, y_{M-k}) - 1 \right| \leq \left| y_{M-k} \right| E[|z_k|] \leq K \left| y_{M-k} \right| \qquad (2.A15)$$

Since $y_{M-k} \to 0$ as $(M-k) \theta \to \infty$, we have $h(k, y_{M-k}) \to 1$ and thus $u(k, y_{N-k}) = h(k, y_{M-k}) v (y_{N-k}) \to 1$ as $\theta(N-k) \to \infty$ which completes the proof.

2.9 Exercises:

2.1) Show that the algorithm given in (2.1) is not distance-diminishing in general, but if $\lambda(p) \equiv a < 1$ and $\psi_i(p) = bP_i$ where $b < 1$ then it is distance-diminishing.

2.2) If $\lambda(p) \equiv a$ and $\psi_i(p) = a\,p_i$ where $a < 1$ show that

$$E[\overset{T}{p}(k+1)] = A\,E[\overset{T}{p}(k)]$$

where $A = [a_{ij}]$ and

$$a_{ii} = a + (1-a)(1-c_i) \quad i = 1,2,\ldots,M$$

$$a_{ij} = \frac{(1-a)c_i}{M-1} \quad \text{for all } i,j, \quad i \neq j$$

Notice A is a stochastic matrix and $a_{ij} > 0$ for all i and j if $0 < c_\ell < 1$ for all $\ell = 1,2,\ldots,M$.

a) Show that

$$\lim_{k \to \infty} E[p_i(k)] = \frac{\dfrac{1}{c_i}}{\displaystyle\sum_{\ell=1}^{M} \frac{1}{c_\ell}}$$

b) Show that the algorithm with the above choice of $\lambda(.)$ and $\psi_i(p)$ is indeed expedient. This is the well known linear reward-penalty algorithm.

2.3) Defining $V[P(k)] \triangleq \mathrm{Var}[P(k)\,|\,P(o) = P]$, we have

$$V[P(k+1)] = V[P(k) + \delta P(k)]$$

$$= V[P(k)] + \mathrm{Var}[\delta P(k)] + 2\,\mathrm{Cov}[P(k),\,\Delta P(k)]$$

where Var [*] is the variance of x and Cov[x,y] is the covariance of x and y. Develop an alternate proof of Corollary 2.3 by estimating the Var[.] and Cov[.,.] terms in the right hand side of $V[P(k+1)]$.

Hint: Refer Lemma 2.2.1 in Norman [N14].

2.4) a. Find the solution of the differential equation

$$\frac{df(t)}{dt} = W[f(t)]$$

where $f(0) = P \in [0,1]$ and $W[P]$ is given by (2.4). Plot the trajectories in the following cases.

    1) $\lambda(P) \equiv 0.1$ and $\psi_1(P) = 0.1P$ and $\psi_2(P) = .1(1-P)$

    2) $\lambda(P) \equiv 0.1$ and $\psi_1(P) = .01P$ and $\psi_2(P) = .01(1-P)$.

  In both cases assume $\theta = 1$, $d_1 = .8$ and $d_2 = .4$.

    b. Simulate the algorithm for the above choices of $\lambda(.)$, $\psi_i(p)$ and parameters noted above and compare the evolution of the "mean learning curve" namely $E[P(k) | P(0) = P]$ with the trajectories obtained in part (a) of this problem.

2.5) Evaluate

$$E[\, |\hat{Z}_\theta(t) - \hat{Z}_\theta(s)|^r \,] \text{ for } r = 1,2$$

where $\hat{Z}_\theta(t)$ is defined in (2.58).

2.6) Consider the Markov diffusion process $\hat{z}(t)$ defined by equation (2.59). Let $\psi(t,y)$ be the transition probability density function of the process $\hat{z}(t)$, where $\hat{z}(0) = 0$. It is well known $\psi(t,y)$ satisfies the following partial differential equation called the Kolmogorov-forward or Fokker-Planck equation:

$$\frac{\partial \, \psi(t,y)}{\partial t} = - W'[f(t)] \frac{\partial [y\psi(t,y)]}{\partial y} + \frac{1}{2} s[f(t)] \frac{\partial^2 \psi(t,y)}{\partial y^2} .$$

    a. Show that the partial differential equation (2.50) can also be obtained from the above Kolmogorov forward or Fokker-Planck equation by Fourier transformation and integration by parts.

(2.7) Under the conditions (C.2.1) - (C2.4), show that the matrix $\dfrac{d\tilde{W}(\tilde{p}^*)}{d\tilde{p}}$ is negative definite.

(2.8)   Complete the proof of Theorem 2.7.

<u>Hint</u>:   Refer Norman [N10] Chapter 8, Theorem 1.1 for a proof of Theorem 2.7

for $k\theta = t < \infty$ and $\theta \to 0$.

CHAPTER 3

ABSOLUTELY EXPEDIENT ALGORITHMS

3.1. Introduction:

This chapter deals with the analysis and design of non-linear reward penalty learning algorithms of the absolutely expedient N_{R-P}^A type. A fundamental property that distinguishes this class of algorithms from those of Chapter 2 is that the set $V_M = \{e_j \mid j = 1,2,\ldots,M, \ e_j = j\text{th unit vector}\}$ consisting of the corners of the simplex S_M form the (only) absorbing states of the Markov process $\{p(k)\}$. Each $\{e_j\} \subset V_M$ is topologically closed and it will be shown that $\{e_j\}$ is also stochastically closed under N_{R-P}^A algorithms. Thus, in this case each element of V_M (there are M of them) according to problem 1.1 form an ergodic kernel as opposed to the only one ergodic kernel (which is S_M) of the N_{R-P}^E algorithms discussed in chapter 2. The presence of such multiple ergodic kernel makes the asymptotic behavior of the Markov process $\{p(k)\}$ under this class of algorithms, as to be expected, very much dependent on the initial state as against the independence of the asymptotic behavior of $\{p(k)\}$ on the initial state in case of the N_{R-P}^E algorithms.

Absolute expediency (refer definition 1.1) consists in making the conditional expectation of $\eta(k+1)$ given the current state $p(k)$ greater than or equal to $\eta(k)$. As $\{p(k)\}$ is a Markov process, this implies that $\eta(k)$ is a non-negative bounded sub-martingale. Using the well known martingale

theorems we first conclude that $\eta(k)$ and hence $p(k)$ converges with probability one. Using the properties of N_{R-P}^A algorithms it is further shown
that $p(k)$ converges to V_M with probability one. In other words, $\lim_{k \to \infty} p(k)$
= $p*$ exists and $p* \varepsilon V_M$ with probability one. This implies in our coin tossing
experiment that asymptotically along each sample path only one coin is
chosen (randomly) for tossing. It is shown that by proper choice of certain
parameters in the algorithm the probability of choosing that coin whose
success probability is maximum can be made as close to unity as desired.
From this as a corollary it is shown that absolute expediency implies
ε - optimality.

A number of examples and simulation results are also presented.

3.2 The N_{R-P}^A algorithm

In our coin tossing experiment at any instant k let coin i be chosen
for tossing as a sample realization from the distribution $p(k)$ =
$(p_1(k), p_2(k) \ldots p_M(k))^T$ where T denotes transpose. We propose to update
$p(k)$ using the following algorithm:

$$p_i(k+1) = p_i(k) + \sum_{j \neq i} \phi_j[p(k)] \left.\begin{array}{l} \end{array}\right\} \quad \text{if coin i is chosen}$$
and the toss

$$p_j(k+1) = p_j(k) - \phi_j[p(k)] \; (j \neq i) \left.\begin{array}{l} \end{array}\right\} \quad \text{resulted in success} \qquad (3.1)$$

$$p_i(k+1) = p_i(k) - \sum_{j \neq i} \psi_j[p(k)] \left.\begin{array}{l} \end{array}\right\} \quad \text{if coin i is chosen}$$
and the toss

$$p_j(k+1) = p_j(k) + \psi_j[p(k)] \; (j \neq i) \left.\begin{array}{l} \end{array}\right\} \quad \text{resulted in failure}$$

where $\phi_j: S_M \rightarrow [0,1]$ and $\psi_j: S_M \rightarrow [0,1]$ are continuous functions for

$j = 1, 2, \ldots, M$ and

(C3.1) Either $\phi_j \equiv 0$ or $0 < \phi_j(p) < p_j, \; \forall \, p_j \in (0,1)$

(C3.2) Either $\psi_j \equiv 0$ or $0 < \sum_{j \neq i} [p_j + \psi_j(p)] < 1, \; \forall \, p_j \in (0,1)$ $\qquad (3.2)$

for all $j = 1, 2, \ldots, M$.

The conditions (C3.1) - (C3.2) ensure that $p(k+1) \; \epsilon \; S_M$ for all $k > 0$

if $p(0)$ does. Further to make the algorithm non-trivial and useful, we only

allow that either $\phi_j(.) \equiv 0$ or $\psi_j(.) \equiv 0$ but not both.

Let

$$\mathcal{D} \triangleq \left\{ D \; \middle| \; \begin{array}{l} D = (d_1, d_2, \ldots, d_m)^T, \; 0 < d_i < 1, \; i=1,2,\ldots,M \\ d_\ell \text{ and } d_m \text{ are unique and } d_\ell > d_m \end{array} \right\} \qquad (3.3)$$

where

$$d_\ell \triangleq \max_j \{d_j\} \text{ and } d_m = \min_j \{d_j\} \qquad (3.4)$$

As in Chapter 2, let

$$\underline{W}(p) = (W_1(p), W_2(p), \ldots, W_M(p))^T \quad \text{where} \left.\begin{array}{l} \\ \end{array}\right\} \qquad (3.5)$$
$$W_i(p) \triangleq E[p_i(k+1) - p_i(k) \mid p(k) = p]$$

It can be seen from (3.1) that

$$W_i(p) = W_i^R(p) + W_i^P(p)$$

$$W_i^R(p) = p_i(k) \, d_i \sum_{j \neq i} \phi_j - \phi_i \sum_{j \neq i} p_j(k) \, d_j \qquad (3.6)$$

$$W_i^P(p) = -p_i(k) \, c_i \sum_{j \neq i} \psi_j + \psi_i \sum_{j \neq i} p_j(k) \, c_j$$

where with little abuse of notation for the sake of simplicity we refer $\phi_j[p(k)]$ by ϕ_j and similarly for ψ_j. Since $p(k) \in S_M$ for all k, as an immediate consequence we have

$$\sum_{i=1}^{M} W_i(p) = 0 \text{ for all } p \in S_M \qquad (3.7)$$

3.3 Conditions for Absolute Expediency:

As a first step toward absolute expediency the following theorem gives conditions under which $W_\ell(p) > 0$ and $W_m(p) < 0$ where the indices ℓ and m are defined in (3.4). Let S_M^o be the interior of S_M defined as

$$S_M^o = \{p \mid \sum_{i=1}^{M} p_i = 1, \; 0 < p_i < 1, \; i = 1,2,3,\ldots,M\} \qquad (3.8)$$

Theorem 3.1: A necessary and sufficient condition for $W_\ell(p) > 0$ and $W_m(p) < 0$ for all $p \in S_M^o$ and $D \in \mathcal{D}$ are

$$(S3.1) \quad \frac{\phi_1}{p_1} = \frac{\phi_2}{p_2} = \ldots\ldots = \frac{\phi_M}{p_M} = \lambda(p)$$

and

$$(S3.2) \quad \frac{\psi_1}{p_1} = \frac{\psi_2}{p_2} = \ldots\ldots = \frac{\psi_M}{p_M} = \mu(p)$$

where $\lambda: S_M \to [0,1]$ and $\mu: S_M \to [0,1]$.

Sufficiency:

Substituting S3.1 and S3.2 in 3.6, on simplification we get

$$W_i(p) = [\phi_i + \psi_i] \sum_{j \neq i} P_j (d_i - d_j)$$

$$= P_i[\lambda(p) + \mu(p)] \sum_{j \neq i} P_j (d_i - d_j) \qquad (3.9)$$

From the properties of the functions $\lambda(.)$ and $\mu(.)$ and from (3.4) we easily

see that $W_\ell(p) > 0$ and $W_m(p) < 0$ for all $p \in S_M^o$ and $D \in \mathcal{D}$.

Necessity: The proof of the necessary part is involved and rather long

and is given in Appendix 3.9   Q.E.D.

Remark 3.1: Conditions S3.1 and S3.2 are called "Symmetry Conditions" and

in view of these conditions C3.1 and C3.2 reduce to

(C' 3.1) $\lambda(p) \equiv 0$ or $0 < \lambda(p) < 1$ for all $p \in S_M^o$

(C' 3.2) $\mu(p) \equiv 0$ or $0 < \mu(p) < \min_j \left\{ \dfrac{P_j}{1-P_j} \right\}$ for $0 < p_j < 1$ (3.10)
$j = 1, 2, \ldots, M$

Typical functions that would be of interest to us are

$$\lambda(p) = a\ p_1^{s_1}\ p_2^{s_2}\ \cdots\ p_M^{s_M}$$

$$\mu(p) = b\ p_1^{r}\ p_2^{r}\ \cdots\ p_M^{r} \qquad (3.11)$$

where constants $s_i \geq 0$ and $r \geq 0$ and a and b are chosen to satisfy (3.10)

Once the signdefiniteness of $W_m(p)$ and $W_\ell(p)$ is assured, absolute

expediency follows immediately as the following theorem demonstrates.

Theorem 3.2: The algorithm (3.1) is absolutely expedient for all $2 \leq M < \infty$

if and only if

$$W_\ell(p) > 0 \quad \text{and} \quad W_m(p) < 0$$

for all $p \in S_M^o$ and $D \in \mathcal{D}$.

Proof Sufficiency: Since theorem 3.1 establishes the equivalence of the

symmetry conditions (S3.1) - (S3.2) and signdefiniteness of $W_\ell(p)$ and

$W_m(p)$, to prove the sufficiency of the latter, it is enough to show that the symmetry conditions imply absolute expediency.

Define

$$\delta\eta(k) \underset{\Delta}{=} E[\eta(k+1) \mid p(k) = p] - \eta(k)$$

From the definition of $\eta(k)$, it follows that

$$\delta\eta(k) = \sum_{i=1}^{M} W_i(p) \, d_i$$

Substituting for $W_i(p)$ from (3.9) on simplification we have (see exercise 3.10)

$$\delta\eta(k) = \frac{1}{2} [\lambda(p) + \mu(p)] \, p^T A \, p \qquad (3.12)$$

where $A = [a_{ij}] = [(d_j - d_i)^2]$ is a real symmetric matrix with all diagonal elements zero and non-diagonal elements positive. Thus far all $d \in \mathcal{D}$ and all $p \notin V_M$, $p^T A \, p > 0$ and $\delta\eta(k) > 0$. This proves that sufficiency.

<u>Necessity</u>: Let $M = 2$ and $d_\ell = d_1$ and $d_m = d_2$. Absolute expediency implies that

$$d_\ell W_\ell(p) + d_m W_m(p) > 0 \text{ for all } p \in s_2^o$$

As $W_\ell(p) + W_m(p) = 0$ for all $p \in S_2$, the above inequality is equivalent to

$$\text{or} \quad \left. \begin{array}{l} W_\ell(p) \, [d_\ell - d_m] > 0 \\[2mm] W_m(p) \, [d_m - d_\ell] > 0 \end{array} \right\} \qquad (3.13)$$

From the definition of d_ℓ and d_m and from (3.13) it immediately follows that $W_\ell(p) > 0$ and $W_m(p) < 0$ for all $p \in s_2^o$.

If $M \geq 3$, absolute expediency requires

$$\sum_{i=1}^{M} W_i(p) \, d_i > 0 \qquad (3.14)$$

for all $D \in \mathcal{D}$. Now consider a specific D in which all d_j's for $j \neq \ell, m$ are equal, that is $d_j = d$ for $j \neq \ell, m$. For this choice of D(3.14) becomes

$$W_\ell(p) \, (d_\ell - d) + W_m(p) \, (d_m - d) > 0 \qquad (3.15)$$

On rewriting

$$W_\ell(p) > \frac{W_m(p)\ (d-d_m)}{(d_\ell - d)} \tag{3.16}$$

As $W_m(p)$ is a bounded function, the right hand side of (3.16) approaches zero as $d \to d_m$. Hence, for (3.16) to hold for all $D \in \mathcal{D}$, $W_\ell(p) > 0$. Similarly by rewriting (3.15) as

$$W_m(p) < \frac{W_\ell(p)\ [d_\ell - d]}{(d - d_m)} \tag{3.17}$$

Again as $W_\ell(p)$ is bounded, as $d \to d_\ell$, it follows from (3.17) that $W_m(p) < 0$.   Q.E.D.

Remark 3.2: A closer look at condition (3.10) reveals that while $\lambda(p) \equiv a$ where $0 < a < 1$ is permitted, the function $\mu(p)$ cannot identically be a constant (other than zero). This explains why the classical L_{R-P}^E algorithm and all the N_{R-P}^E algorithms discussed in chapter 2 are not absolutely expedient (See exercise 3.1).

As an immediate consequence we have the following:

Corollary 3.1: Absolute expediency implies expediency, if $p(0) = (\frac{1}{M}, \frac{1}{M}, \cdots, \frac{1}{M})^T$.

Proof: Refer footnote in section 1.3.   Q.E.D.

Remark 3.3: The converse of Corollary 3.1 is not true, however, for the L_{R-P}^E learning algorithm is expedient but not absolutely expedient. If $\lambda(p) \not\equiv 0$ but $\mu(p) \equiv 0$ is known as non-linear reward-inaction algorithm - N_{R-I}^A. If $\lambda(p) \equiv a$, a constant where $0 < a < 1$ and $\mu(p) \equiv 0$, we get the linear reward-inaction L_{R-I}^A algorithm. It easily follows that N_{R-I}^A, L_{R-I}^A are both absolutely expedient. (see exercise 3.11).

Using the symmetry conditions we rewrite the algorithm (3.1) as

$$p_i(k+1) = p_i(k) + \lambda[p(k)] [1 - p_i(k)]$$

$$p_j(k+1) = p_j(k) - \lambda[p(k)] \, p_j(k), \quad (j \neq i)$$

if coin i is chosen and toss resulted in success

(3.1a)

$$p_i(k+1) = p_i(k) - \mu[p(k)] [1 - p_i(k)]$$

$$p_j(k+1) = p_j(k) + \mu[p(k)] \, p_j(k) \quad (j \neq i)$$

if coin i is chosen and toss resulted in failure

or in the vector form as

$$p(k+1) = p(k) + [e_i - p(k)] \, \lambda[p(k)]$$

if coin i is chosen and toss resulted in success

$$p(k+1) = p(k) - [e_i - p(k)] \, \mu[p(k)]$$

if coin i is chosen and toss resulted in failure

(3.1b)

3.4 Analysis of Absolutely Expedient Algorithms:

So far our attention has been confined to finding conditions under which the N_{R-P}^A algorithm is absolutely expedient. As a consequence we have also shown that the N_{R-P}^A algorithm is expedient. In this section we set out to analyze the actual sample path behavior of the process $\{P(k)\}$ that evolve according to algorithm (3.1b). To this end we first need an addition condition on $\lambda(P)$:

(C'3.3) $\lambda(p) = 0$ only if $p \in V_M$.

Notice that this condition ensures that $\lambda(p)$ cannot be zero unless p is a unit vector and does not exclude cases when $\lambda(p)$ is not zero for any other value of the argument.

Define $\mathcal{D}_1 \subset \mathcal{D}$ as

$$\mathcal{D}_1 = \{D \mid D = (d_1, d_2, \ldots d_M)^T, \ 0 < d_i < 1, \ d_i \neq d_j \text{ for all } i \neq j, \ i, j = 1, 2, 3, \ldots, M\}$$

(3.18)

Notice if $D \in \mathcal{D}_1$ as the components of D are distinct, the existence of unique d_ℓ and d_m satisfying (3.4) automatically follows.

Our next theorem characterizes an important property of the set V_M with respect to the algorithm (3.1b).

Theorem 3.3: If $\lambda(.)$ and $\mu(.)$ satisfy (C'3.1) – (C'3.3) then for all $D \in \mathcal{D}_1$, the set of all unit vectors (namely the set of all elements of V_M) form the only absorbing barriers of the Markov process $\{p(k)\}$ generated by (3.1b)

Proof: From (3.1b) and from the conditions of the theorem 3.3, it can be seen that $p(k+1) = p(k)$ with probability one if and only if $p(k) \in V_M$. We invite the reader to fill in the details. (See exercise 3.2).   Q.E.D.

Thus V_M constitutes the invariant set or the set of all absorbing barriers for the process $p(k)$. Now we are ready to state a fundamental result on the sample path behavior of the process $\{p(k)\}$.

Theorem 3.4: Under conditions (C'3.1)-(C'3.3) and for all $D \in \mathcal{D}_1$ the Markov process $\{p(k)\}$ that evolves according to algorithm (3.1a and b) converges to the set V_M with probability one.

Proof: Let $F[x(1),x(2),\ldots, X(k)]$ be the σ-algebra generated by a random sequence $x(1), x(2), \ldots, x(k)$. As $\eta(k)$ is a linear function of $p(k)$, it easily follows that [T 7]

$$F[\eta(1), \eta(2), \ldots, \eta(k)] \subset F[p(1), p(2), \ldots, p(k)] \qquad (3.19)$$

Further

$$F[p(k)] \subset F[p(1),p(2),\ldots p(k)] \qquad (3.20)$$

From theorem (3.2) we know that

$$\left. \begin{array}{l} \delta\eta(k) = E[\ \eta(k+1) \mid p(k)] - \eta(k) \\ \qquad = \dfrac{1}{2} [\lambda(p) + \mu(p)] \ p^T A \ p \end{array} \right\} \qquad (3.12)$$

where $A = [a_{ij}] = [(d_j - d_i)^2]$

Notice that for all $p \in S_M$ and for all $D \in \mathcal{D}_1$, the right hand side of (3.12) is non-negative and hence using (3.20) we get

$$E[\eta (k+1) \mid F[p(1), p(2), \ldots, p(k)]] \geq \eta(k) \tag{3.21}$$

Now taking conditional expectations on both sides of (3.12) with respect to $F[\eta(1), \eta(2), \ldots, \eta(k)]$ from the law of iterated conditional expectations [T 7] [D2] and from (3.19) we get

$$E[\eta(k+1) \mid F[\eta(1), \eta(2), \ldots, \eta(k)]] \geq \eta(k) \tag{3.22}$$

This in turn implies [T7] [D2] that $\eta(k)$ is a non-negative bounded sub-martingale and hence $\eta(k)$ converges to a random variable η^* with probability one. As $\eta(k)$ is uniformly bounded (from sub-martingale theorem [T 7] [D2] we have $E[\eta(k)]$ also converges to $E[\eta^*]$. Consequently $\lim\limits_{k \to \infty} E[\delta\eta(k)]$ must vanish. As $\delta\eta(k)$ is a non-negative random variable it follows that $\lim\limits_{k \to \infty} \delta\eta(k)$ must vanish with probability one. For all $D \in \mathcal{D}_1$ it easily follows from (3.12) and (C'3.3) that $\lim\limits_{k \to \infty} \delta\eta(k) = 0$ if and only if

$$\lim\limits_{k \to \infty} p(k) = p^* \in V_M \text{ with probability one} \qquad \text{Q.E.D.}$$

Recall that if $p(k) = e_j$ then $p(k+1) = e_j$ with probability one. Hence $\{e_j\}$ as a subset of V_M is stochastically closed. Also as $\{e_j\}$ is topologically closed, each element of V_M, as a set, constitutes an ergodic kernel.

The above theorem brings out the intrinsic difference between Ergodic algorithms of Chapter 2 and the absolutely expedient algorithms which is the topic of this chapter. Recall that in Chapter 2 we showed that the distribution of $p(k)$ after suitable normalization is Gaussian. In other words in the case of Ergodic algorithms $p(k)$ converges in distribution. But $p(k)$ when generated by an absolutely expedient algorithm converges with probability one to a random vector whose range is a discrete, finite set. Our aim in the following is to characterize the discrete distribution of p^*.

Define

$$\Gamma_i(p) = \text{Prob } [\ p^* = e_i \ | \ p(0) = p] \quad i = 1,2,\ldots,M$$
$$= \text{Prob } [\ p_i^* = 1 \ | \ p(0) = p] \tag{3.22}$$

Notice $\sum_{i=1}^{M} \Gamma_i(p) = 1$ for all $p \in S_M$. Stated in words, the above theorem 3.4 states that in our coin tossing experiment if we update $p(k)$ according to (3.1b) then along each sample path asymptotically we will end up choosing only one coin and coin i will be chosen with probability $\Gamma_i(p)$ where $p(0)=p$.

In much of what follows in this section we will characterize further properties of $\Gamma_i(p)$. To this end we first record few useful definitions.

Let $C[S_M]$ be the class of all continuous functions from S_M to the real line. If $f(.) \in C[S_M]$, then define the operator u as

$$\left. \begin{array}{l} uf(p) = E[f \ [p(k+1) \ ] \ | \ p(k) = p] \\[2mm] \qquad = \displaystyle\int_{S_M} f(x) \ K[p,dx] \end{array} \right\} \tag{3.23}$$

where $K[.,.]$ is the transition function defined as

$$K[p,A] = \text{Prob } [p(k+1) \in A \ | \ p(k) = p] \quad \forall \ k \geq 0 \tag{3.24}$$

where A is a (Borel) subset of S_M. In our case the transition function is defined in terms of the algorithm (3.1b). Thus, as an example

$$uf(p) = p_i[d_i \ f[p+(e_i-p) \ \lambda(p)] + c_i \ f[p-(e_i-p) \ \mu(p)] \ ]$$
$$+ \sum_{j \neq 1} p_j[d_j \ f[p + (e_j-p) \ \lambda(p)] + c_j f \ [p - (e_i - p) \ \mu(p)] \tag{3.25}$$

Define

$$K^{(n)}[p,A] = \text{Prob } [p(k+n) \in A \ | \ p(k) = p] \tag{3.26}$$

for $n \geq 1$ and all $k \geq 0$, as the n-step transition function. The following are immediate consequence of the definition of u.

1. u is <u>linear</u>. That is, if $f_1(p), f_2(p) \in C[S_M]$ and m_1, m_2 are real constants then

$$u\,(m_1\, f_1\, [p] + m_2\, f_2\, [p]) = m_1\, u\, f_1\, [p] + m_2\, u\, f_2(p).$$

2. u is positive, that is, it preserves non-negative functions. Thus, if $f(p)$ is non-negative so is $uf(p)$.

3.
$$u^2\, f(p) = u\, [u\, f\, [p]\,]$$

$$= E\, [u\, f\, [p(k+1)]\, \,|p(k) = p]$$

$$= E\, [E\, [f\, [p(k+2)\,]\,|\, p(k+1)\,]\,|\, p(k) = p]$$

$$= \int_{S_M} K[p, dx]\, \int_{S_M} K[x,\, dy]\, f[y]$$

$$= \int_{S_M} K^{(2)}\, [p,\, dy]\, f(y) = E[f\, [p(k+2)]\, |\, p(k) = p\,]$$

where $K^{(2)}\, [p, A] = \int_{S_M} K[p, dx]\, K[x, A]$ is the well known Chapman-Kolmogorov equation [D2]. In general for any $n \geq 2$

$$u^n\, f[p] = E[\, f[p(k+n)]\, |\, p(k) = p]$$

$$= \int_{S_M} K^{(n)}\, [p,\, dx]\, f(x)$$

where $K^{(n)}\, [p, \Lambda] = \int_{S_M} K[p,\, dx]\, K^{(n-1)}\, [x, A]$ and u^n is called the nth iterate of u.

<u>Definition 3.1</u>: A function $f: S_M \to R$, the real line is called super regular (regular, sub regular) function if and only if

$$f(p) \geq (\,=, \leq)\, uf(p)\, \forall\, p \in S_M \tag{3.26}$$

Remark 3.4: It easily follows from the above definition that constant
functions are regular and hence are both sub and super regular. Also if f(.)
is a subregular function then -f(.) is super regular. Further if $f_1(.)$ and
$f_2(.)$ are super (sub) regular functions and if m_1 and m_2 are real, positive
constants, then $m_1 f_1(.) + m_2 f_2(.)$ is also super (sub) regular. In other
words, the class of super (sub) regular functions are closed under addition
and multiplication by positive constants.

Example 3.1: Let $f_\ell(p) = p_\ell$ and $f_m(p) = p_m$. From the definition it follows that

$$u\, f_\ell(p) - f_\ell(p) = W_\ell(p) > 0$$

$$u\, f_m(p) - f_m(p) = W_m(p) < 0.$$

Thus $f_\ell(p)$ and $f_m(p)$ are sub and super regular functions respectively.

With these preliminaries we now state and prove two propositions which
suggest an algorithm to quantify $\Gamma_i(p)$.

Proposition 3.1: $\Gamma_i(p)$ is the only continuous solution of the functional
equation

$$u\, \Gamma_i(p) = \Gamma_i(p) \tag{3.27}$$

satisfying the boundary condition

$$\Gamma_i(e_i) = 1 \text{ and } \Gamma_i(e_j) = 0 \text{ for } i \neq j \tag{3.28}$$

The proof of this important proposition is given in Appendix 3.9.
Notice $\Gamma_i(p)$ satisfying (3.27) by definition is a regular function. The
proposition at least in principle completely solves the problem of
characterizing $\Gamma_i(p)$. However, it may be surprising to note that except
for certain very simple cases [N11] no one has yet been able to solve the
above functional equation. Discomforting as it may seem, the following
proposition provides some relief.

Proposition 3.2: Let $f_i(p) \in C[S_M]$ be super (sub) regular with $f_i(e_i) = 1$
and $f_i(e_j) = 0$ for $j \neq i$. Then

$$f_i(p) \geq (\leq) \; \Gamma_i(p) \tag{3.29}$$

Proof: Let $f_i(p)$ be super regular, that is,

$$f_i(p) \geq u \; f_i(p)$$

Since u is positive, we have

$$u \; f_i(p) \geq u^2 \; f_i(p) \geq \ldots \geq u^\infty \; f_i(p)$$

But as $\lim_{k \to \infty} p(k) = p^*$,

$$u^\infty \; f_i(p) = E \; [f_i(p^*) \mid p(0) = p]$$

$$= \sum_{j=1}^{M} f_i(e_j) \; \Gamma_j(p)$$

$$= \Gamma_i(p)$$

Hence $f_i(p) \geq \Gamma_i(p)$

The result for sub-regular function follows in a similar manner.   Q.E.D.

Thus, if we can find two functions $h_i^{(1)}(p)$ and $h_i^{(2)}(p)$ which are super and sub regular functions respectively and satisfying the boundary conditions

$$h_i^{(1)}(e_i) = h_i^{(2)}(e_i) = 1$$

$$h_i^{(1)}(e_j) = h_i^{(2)}(e_j) = 0 \quad \text{for } j \neq i \tag{3.30}$$

Then from the above proposition it follows that

$$h_i^{(2)}(p) \leq \Gamma_i(p) \leq h_i^{(1)}(p) \tag{3.31}$$

In other words any super and sub regular functions satisfying (3.30) form the upper and lower bounds on $\Gamma_i(p)$. These bounds are relatively easy to compute and in the following section we provide an algorithm to compute these bounds.

3.5 An Algorithm to Compute the Bounds:

Consider a function

$$\psi^i[x_i,p] = e^{-x_i p_i} \tag{3.32}$$

where $x_i > 0$ is a parameter

clearly $1 = \psi^i[x_i,e_j] < \psi^i[x_i,e_i] = e^{-x_i}$ for $i \neq j$

Define $\phi^i[x_i p] = \dfrac{1-e^{-x_i p_i}}{1-e^{-x_i}}$ for $x_i > 0$ \tag{3.33}

From remark 3.4 it follows that if $\psi^i[x_i,p]$ is sub (super) regular, then $\phi^i[x_i,p]$ is super (sub) regular.

Also it easily follows from (3.33) that

$$\phi^i[x_i,e_i] = 1 \text{ and } \phi^i[x_i,e_j] = 0 \tag{3.34}$$

Let y_i and z_i be two positive constants such that

$$\left. \begin{array}{l} \phi^i[y_i p] = \dfrac{1-e^{-y_i p_i}}{1-e^{-y_i}} \text{ is sub-regular} \\[4mm] \phi^i[z_i,p] = \dfrac{1-e^{-z_i p_i}}{1-e^{-z_i}} \text{ is super-regular} \end{array} \right\} \tag{3.35}$$

and

Combining (3.34) and (3.35), in view of (3.30) and (3.31), we can sandwich $\Gamma_i(p)$ as

$$\frac{1-e^{-y_i p_i}}{1-e^{-y_i}} = \phi^i[y_i,p] \leq \Gamma_i(p) \leq \phi^i[z_i p] = \frac{1-e^{-z_i p_i}}{1-e^{-z_i}} \tag{3.36}$$

From this discussion it follows that the problem obtaining (lower and upper) bound on $\Gamma_i(p)$ reduces to one of finding two constants $y_i > 0$ and $z_i > 0$ such that (3.35) is true.

To this end let us first calculate $u \, e^{-x_i p_i}$. It can be seen from (3.25) that

$$u \, e^{-x_i p_i} - e^{-x_i p_i} = -x_i \, F_i \, [x_i, p] e^{-x_i p_i} \qquad (3.37)$$

where

$$F_i[x_i p] = p_i(1-p_i) \, d_i \, \lambda(p) \, V[-x_i(1-p_i) \, \lambda(p)]$$

$$-p_i(1-p_i) \, c_i \, \mu(p) \, V[\, x_i(1-p_i) \, \mu(p)]$$

$$-(\sum_{j \neq i} p_j \, d_j \,) \, p_i \, \lambda(p) \, V[x_i \, p_i \, \lambda(p)]$$

$$+(\sum_{j \neq i} p_j \, c_j \,) \, p_i \, \mu(p) \, V[-x_i \, p_i \, \mu(p)] \qquad (3.38)$$

and

$$V[u] = \begin{cases} \dfrac{e^u - 1}{u} & \text{for } u \neq 0 \\ 1 & \text{for } u = 0 \end{cases} \qquad (3.39)$$

If $y_i > 0$ and $z_i > 0$ are such that

$$F[y_i, p] \geq 0 \quad \text{and} \quad F[z_i, p] \leq 0 \qquad (3.39)$$

then from (3.37), (3.32) and (3.33) we easily see that

$\psi^i[y_i, p]$ is super regular and $\psi^i[z_i, p]$ is sub regular

and

$\phi^i[y_i, p]$ is sub regular and $\phi^i[z_i, p]$ is super regular $\qquad (3.40)$

Before presenting a procedure for finding y_i and z_i satisfying (3.39) we shall record some of the properties of the function $V[.]$ for our later use.

(a) $V[u] > 0$ for all u

(b) $V[u]$ is strictly monotonically increasing $\qquad (3.41)$

(c) $V[u]$ is convex

If we define

$$H[u] = \log_e V[u] \qquad (3.42)$$

it follows that

    d) H[u] is strictly monotonically increasing

    e) H[u] is convex

$$\hspace{8cm}(3.43)$$

For the purpose of illustrating the idea, a special but not trivial case of the N_{R-I}^{A} algorithm is considered first by setting $\mu(.) \equiv 0$ (refer remark 3.3)

Case 1. N_{R-I}^{A} algorithm:

    Setting $\mu(.) \equiv 0$ in (3.38) it follows that $\psi^{i}[x_{i},p]$ is sub (super) regular if

$$G[x_{i}p] \triangleq \frac{V[-x_{i}(1-p_{i})\lambda(p)]}{V[x_{i} \ p_{i} \ \lambda(p) \ ]} \underset{(\geq)}{\leq} \frac{\sum\limits_{j \neq i} p_{j} d_{j}}{(1-p_{i}) \ d_{i}} \hspace{3cm}(3.44)$$

The right hand side of (3.44) can be written as

$$\frac{\sum\limits_{j \neq i} p_{j} d_{j}}{(1-p_{i}) d_{i}} = \frac{1}{\left(\sum\limits_{j \neq i} p_{j}\right)} \ \sum\limits_{j \neq i} p_{j} \ \left(\frac{d_{j}}{d_{i}}\right)$$

from this we see that

$$\min_{j \neq i} \ \left(\frac{d_{j}}{d_{i}}\right) \leq \frac{1}{\left(\sum\limits_{j \neq i} p_{j}\right)} \ \sum\limits_{j \neq i} p_{j} \ \left(\frac{d_{j}}{d_{i}}\right) \leq \max_{j \neq i} \ \left(\frac{d_{j}}{d_{i}}\right) \hspace{2cm}(3.45)$$

Combining (3.44) and (3.45) it follows that

    $\psi^{i}[x_{i},p]$ is sub regular if

$$G[x_{i},p] \leq \min_{j \neq i} \ \left(\frac{d_{j}}{d_{i}}\right) \hspace{5cm}(3.46)$$

and

    $\psi^{i}[x_{i},b]$ is super regular if

$$G[x_{i},p] \geq \max_{j \neq i} \ \left(\frac{d_{j}}{d_{i}}\right) \hspace{5cm}(3.47)$$

for all $p \ \varepsilon \ S_{M}$

Define

$$\Delta[x_i, p] = \log_e G[x_i, p] \tag{3.48}$$

From (3.42) and (3.44) we get

$$\Delta[x_i, p] = H[-x_i (1-p_i) \lambda(p)] - H[x_i p_i \lambda(p)]$$

$$= \int_{x_i p_i \lambda(p)}^{-x_i(1-p_i)\lambda(p)} \left(\frac{d H(u)}{du}\right) du \tag{3.49}$$

Since $H[u]$ is non-negative and strictly monotonically increasing, it can be shown (Exercise 3.6) that

$$g_1[x_i] \leq \Delta[x_i, p] \leq g_2[x_i] \tag{3.50}$$

where

$$\left.\begin{array}{l} g_1[x_i] = -\displaystyle\int_0^{x_i \lambda_{max}} \left[\frac{d H[u]}{du}\right] du \\[2em] g_2[x_i] = -\displaystyle\int_{-x_i\lambda_{min}}^{0} \left[\frac{dH[u]}{du}\right] du \end{array}\right\} \tag{3.51}$$

$$\lambda_{min} = \inf_{p \varepsilon S_M} \lambda(p) \quad \text{and} \quad \lambda_{max} = \sup_{p \varepsilon S_M} \lambda(p) \tag{3.52}$$

Taking exponentials throughout (3.50) we obtain

$$\frac{1}{V[x_i \lambda_{max}]} \leq G[x_i, p] \leq V[-x_i \lambda_{min}] \tag{3.53}$$

Setting

$$V[-x_i \lambda_{min}] = \min_{j \neq i} \left\{\frac{d_j}{d_i}\right\} \tag{3.54}$$

let $Z_i > 0$ be the value of x_i for which (3.54) is true.

Similarly setting

$$V[x_i \; \lambda_{max}] = \cfrac{1}{\max\limits_{j \neq i} \{ \cfrac{d_j}{d_i} \}} \tag{3.55}$$

let $y_i > 0$ be the value of x_i for which (3.55) is true.

Now for all $D \in \mathcal{D}_1$ and from the definition of d_ℓ (in 3.4) we see that

for $i = \ell$

$$\min\limits_{j \neq \ell} \{ \frac{d_j}{d_\ell} \} < \max\limits_{j \neq \ell} \{ \frac{d_j}{d_\ell} \} < 1 \tag{3.56}$$

As

$$V[u] \begin{cases} < 1 & \text{for } u < 0 \\ \\ > 1 & \text{for } u > 0 \end{cases} \tag{3.56a}$$

and $V[u]$ is strictly monotonically increasing from (3.54) - (3.56a) it follows

that there exists unique values for z_ℓ and y_ℓ with the above said properties.

Algorithm for finding y_ℓ and z_ℓ:

STEP 1: Calculate λ_{min} and λ_{max}

STEP 2: Compute $\max\limits_{j \neq \ell} \{ \frac{d_j}{d_\ell} \}$ and $\min\limits_{j \neq \ell} \{ \frac{d_j}{d_\ell} \}$

STEP 3: Solve equations (3.54) and (3.55) for $z_\ell > 0$ and $y_\ell > 0$.

Remark 3.5: Substituting the values of z_ℓ and y_ℓ in (3.36) (with $i = \ell$) we

easily obtain both the lower and upper bounds on $\Gamma_\ell(p)$. The reason for

singling out the index ℓ is that it is the index of the coin with the largest

probability of success. Hence $\Gamma_\ell(p)$ refers to the asymptotic probability

with which the coin with maximum probability of success will be chosen for

tossing along each sample path. It is intuitively clear that if we can make

$\Gamma_\ell(p)$ very close to unity we can hope to achieve ε-optimality. It will be shown in the following section that such an intuition is indeed correct.

Remark 3.6:

Only for simplicity in presentation did we choose the function $\psi^i[x_i,p]$
in (3.32) to depend on only one component, p_i only. By considering any
other functions which depend on all components of p we may obtain better
(that is, tighter) bounds, at the expense of rather complex and involved
algebra. In the special case of M = 2, even with such a simple choice of
function as in (3.32) it is shown below that much tighter bounds are possible.

Special Case of M = 2: We shall illustrate our claim by deriving a tighter
lower bound for $\Delta[x_i,p]$ defined in (3.48). Without loss of generality let
$D = (d_\ell, d_k)$ and $p = (p_\ell, p_k)$ where $0 < p_\ell < 1$ and $p_\ell + p_k = 1$. Rewrite (3.49)
as (with i = ℓ)

$$-\Delta[x_\ell, p] = \int_{-x_\ell(1-p_\ell)\lambda(p)}^{0} [\frac{d\ H[u]}{du}]du + \int_{0}^{x_\ell p_\ell \lambda(p)} [\frac{dH[u]}{du}]\ du \qquad (3.57)$$

It can be shown (see exercise 3.7) that

$$\frac{dH[u]}{du} < \frac{1}{2} \quad \text{for } u < 0$$
$$\frac{dH[u]}{du} < [1 - \frac{1}{2}e^{-0.2u}] \quad \text{for } u > 0 \qquad (3.58)$$

Substituting (3.58) in (3.57)

$$-\Delta[x_\ell,p] \le \frac{1}{2}\ x_\ell(1-p_\ell)\lambda(p) + \int_{0}^{x_\ell p_\ell \lambda(p)} [1 - \frac{1}{2}e^{-0.2u}]\ du$$

$$= h_\ell[x_\ell,p] \quad \text{say} \qquad (3.59)$$

Define

$$g_3[x_\ell] = \inf_{0<p_\ell<1} \{-h_\ell[x_\ell,p]\} \qquad (3.60)$$

Combining (3.57) and (3.60) we can rewrite (3.50) as (with i = ℓ)

$$g_3[x_\ell] \leq \Delta[x_\ell, p] \leq g_2[x_\ell] \tag{3.61}$$

Taking exponentials throughout (3.61) we get

$$e^{g_3[x_\ell]} \leq G[x_\ell, p] \leq V[-x_\ell \lambda_{min}] \tag{3.62}$$

Notice in this special case

$$\max_{j \neq \ell} \{ \frac{d_j}{d_\ell} \} = \min_{j \neq \ell} \{ \frac{d_j}{d_\ell} \} = \frac{d_k}{d_\ell} < 1 \tag{3.63}$$

Let z_ℓ and y_ℓ be defined by

$$V[-z_\ell \lambda_{min}] = \frac{d_k}{d_\ell}$$

and

$$\left. e^{g_3[y_\ell]} = \frac{d_k}{d_\ell} \right\} \tag{3.64}$$

As $h_\ell[0,p] = 0$ and $h_\ell[x_\ell,p] \to \infty$ when $x_\ell \to \infty$, we have $g_3[x_\ell]$ is non-positive for all x_ℓ. From this and the properties of $V[.]$ it follows that z_ℓ and y_ℓ satisfying (3.64) exist.

In addition to $M = 2$ if $\lambda(p) \equiv a$, a constant $0 < a < 1$ and $\mu(p) \equiv 0$; then as $\lambda_{min} = \lambda_{max} = a$ from (3.54), (3.55) and (3.64) we have

$$\left. \begin{array}{l} V[-z_\ell a] = \frac{d_k}{d_\ell} < 1 \\[2em] V[-y_\ell a] = \frac{d_\ell}{d_k} > 1 \end{array} \right\} \tag{3.65}$$

Example 3.2: Consider an example with $M = 4$ and the vector of success probabilities be $D = (0.90, 0.55, .40, 0.05)$

$$d_\ell = 0.90; \quad d_k = 0.05; \quad \max_{j \neq \ell} (\frac{d_j}{d_\ell}) = \frac{0.55}{0.90} = \frac{11}{18}; \quad \min_{j \neq \ell} (\frac{d_j}{d_\ell}) = \frac{0.05}{0.90} = \frac{1}{18}$$

Let $\lambda(p) \equiv 0.10$ and $\mu(p) \equiv 0$

From (3.54) and (3.55) the following computations are easily verified.

$$V[-z_\ell \, \lambda_{min}] = V[-0.10 \, z_\ell] = \frac{1}{18}$$

$$V[y_\ell \, \lambda_{max}] = V[0.1 \, y_\ell] = \frac{18}{11}$$

On solving the last two equations we get

$y_\ell \simeq 9.15$ and $z_\ell \simeq 180$

For various values of p_ℓ the lower and upper bounds on $\Gamma_\ell(p)$ are given in table 3.1.

Table 3.1

P_ℓ	Bounds on $\Gamma_\ell(p)$	
	Lower Bound	Upper bound
.1	0.5995	1.00
.2	0.8396	1.00
.3	0.9358	1.00
.4	0.9744	1.00
.5	0.9898	1.00
.6	0.9959	1.00
.7	0.9985	1.00
.8	≈ 1.0	1.00
.9	≈ 1.0	1.00

Case 2: N_{R-P}^A algorithm ($\lambda(.) \neq 0$, $\mu(.) \neq 0$);

In this general case it is not possible to give explicit equations for determining y_ℓ and z_ℓ as in the N_{R-I}^A algorithm. However, the procedure is quite straightforward and is as follows.

Define

$$\left. \begin{aligned} g_3(x_\ell) &= \sup_{p \, S_M^o} \, F_\ell[x_\ell, p] \\[2ex] g_4(x_\ell) &= \inf_{p \, S_M^o} \, F_\ell[x_\ell, p] \end{aligned} \right\} \tag{3.66}$$

Let $y_\ell > 0$ be the largest value of x_ℓ for which

$$g_4(x_\ell) \Big|_{x_\ell = y_\ell} \geq 0 \qquad\qquad (3.67)$$

and let $z_i > 0$ be the smallest value of x_ℓ for which

$$g_3(x_\ell) \Big|_{x_\ell = z_\ell} \leq 0 \qquad\qquad (3.68)$$

From (3.38) it can be seen, for all $D \in \mathcal{D}_1$ and $p \in S_M^0$ that

(a) $F_\ell(0,p) = p_\ell \, \lambda(p) \sum\limits_{j \neq \ell} P_j \, (d_\ell - d_j)$

$$+ \, p_\ell \, \mu(p) \sum\limits_{j \neq \ell} P_j \, (c_j - c_\ell)$$

$$> 0$$

(b) as $V[u] \to \infty$ as $u \to \infty$ and $v[u] \to 0$ as $u \to -\infty$ it follows from (3.38)

$$\lim_{x_i \to \infty} F_\ell[x_\ell, p] \to -\infty$$

Hence, $y_\ell > 0$ and $z_\ell > 0$ satisfying (3.67) and (3.68) exist. Thus, at least in principle we can find the bounds on $\Gamma_\ell(p)$.

Example 3.3: $M = 2$; $D = (.8, .4)$; $d_\ell = 0.8$; $d_k = 0.4$ $d_k/d_\ell = \frac{1}{2}$. Two algorithms are considered.

L_{R-I}^A : $\lambda(p) \equiv 0.08$; $\mu(p) \equiv 00$

N_{R-P}^A : $\lambda(p) = 0.05 + 0.05 \, p_1 p_2$ $\mu(p) = 0.1 \, p_1 \, p_2$

using the method illustrated above, the following results are obtained.

	y_ℓ	z_ℓ
L_{R-1}^A	15.75	20.5

The upper and lower bounds on $\Gamma_\ell(p)$ for both the algorithms are tabulated below.

Table 3.2 L_{R-I}^{A}: $\lambda(p) \equiv 0.08$; $\mu(p) \equiv 0$

P_ℓ	Bounds on $\Gamma_\ell(p)$	
	Lower bound	Upper bound
.1	0.7930	.8713
.2	0.9572	.9834
.3	0.9911	.9979
.4	0.9982	.9999
.5	0.9996	1.0
.6	$\approx$ 1.0	1.0

Table 3.3 N_{R-P}^{A}: $\lambda(p) = 0.05 + 05\ p_1 p_2$; $\mu(p) = 0.1\ p_1 p_2$

P_ℓ	Bounds on $\Gamma_\ell(p)$	
	Lower bound	Upper bound
0.1	0.9219	0.9740
0.2	0.9939	0.9993
0.3	0.9995	1.0
0.4	1.0	1.0
0.5	1.0	1.0

3.6 Absolute Expediency and ε-optimality:

In corollary 3.1 we showed that absolute expediency implies expediency. In this section using the bounds on $\Gamma_\ell(p)$ it will be shown absolute expediency indeed is a sufficient condition for ε-optimality. To this end let us rewrite

$$\left.\begin{array}{l} \lambda(p) = \theta\lambda_1(p) \\[2mm] \mu(p) = \theta\mu_1(p) \end{array}\right\} \tag{3.69}$$

where θ is real and $0 < \theta < 1$ and $\lambda_1(.)$ and $\mu_1(.)$ satisfy the conditions $(C'3.1) - (C'3.3)$. Clearly, introduction of the parameter θ enables us to control the step-length in the evolution of the Markov process $\{p(k)\}$ defined in (3.13). In particular our interest is in the investigation of the effect of small θ, that is, $\theta \approx 0$.

For small values of u, using the third order expansion for e^u we can write

$$V[u] \approx 1 + \frac{u}{2} \tag{3.70}$$

Substituting (3.69) in (3.38) and using (3.70), we can see, after simplification

$$F[x_i,p] = \theta[\lambda_1(p) + \mu_1(p)] \; p_i \sum_{j \neq i} p_j(d_i - d_j) - \frac{1}{2}\theta^2 \; x_i \; B(p) \tag{3.71}$$

where

$$\begin{aligned} B(p) = \; & p_i \; \lambda_1^2(p) \; [\; (1-p_i)^2 \; d_i + p_i \sum_{j \neq i} p_j \; d_j] \\ & + p_i \; \mu_1^2(p) \; [(1-p_i)^2 \; c_i + p_i \sum_{j \neq i} p_j \; c_j] \end{aligned} \tag{3.72}$$

For all $p \in S_M^o$, clearly $B(p) > 0$. Hence $\psi_1[x_{i,p}]$ is super (sub) regular if

$$x_i \begin{array}{c} \geq \\ (\leq) \end{array} \frac{[\lambda_1(p) + \mu_1(p)] \sum\limits_{j \neq i} p_j(d_i - d_j)}{\frac{1}{2} \theta \; B(p)} \tag{3.73}$$

If $i = \ell$ recall that $d_\ell - d_j > 0$ for all $j \neq \ell$. Thus for $i = \ell$, the right hand side of (3.73) is positive and $\to \infty$ as $\theta \to 0$. Hence, y_ℓ in (3.36) (which is the largest value of x_ℓ satisfying the second inequality $(\leq)$ in (3.73)) approaches $\to \infty$ as $\theta \to 0$.

From this discussion it follows that for all $p \in S_M^o$ from (3.36)

$$\Gamma_\ell(p) \geq \frac{1 - e^{-y_\ell p_\ell}}{1 - e^{-y_\ell}} \to 1 \text{ as } \theta \to 0 \qquad (3.74)$$

Now let us compute $\lim_{k \to \infty} E[\eta(k)]$

$$\lim_{k \to \infty} E[\eta(k)] = \lim_{k \to \infty} \sum_{j=1}^{M} E[p_j(k) \, d_j]$$

$$= \sum_{j=1}^{M} E[\lim_{k \to \infty} p_j(k)] \, d_j$$

$$= \sum_{j=1}^{M} \Gamma_j(p) \, d_j \quad \text{by theorem (3.4)}$$

$$= d_\ell - \sum_{j \neq \ell} \Gamma_j(p) \, [d_\ell - d_j]$$

Therefore

$$\lim_{k \to \infty} \left| E[\eta(k) - d_\ell] \right| \leq \sum_{j \neq \ell} \Gamma_j(p) \, [d_\ell - d_j]$$

$$< (d_\ell - d_k) \, [1 - \Gamma_\ell(p)] \qquad (3.75)$$

Given any $\delta > 0$ from (3.74) it follows that there exists a $0 < \theta^*$ such that for all $\theta < \theta^*$

$$\Gamma_\ell(p) \geq (1 - \delta) \qquad (3.76)$$

Combining (3.75) and (3.76) we get for all $0 < \theta < \theta^*$

$$\lim_{k \to \infty} \left| E[\eta(k) - d_\ell] \right| < (d_\ell - d_k) \, \delta < \varepsilon \qquad (3.76)$$

for $\delta < \dfrac{\varepsilon}{(d_\ell - d_k)}$. The following theorem summarizes the above discussion.

Theorem 3.5: There exists proper choice of functions $\lambda(.)$ and $\mu(.)$ satis-fying the conditions $(C'3.1) - (C'3.3)$ such that the absolutely expedient algorithm is ε-optimal.

3.7   SIMULATIONS

In this section we present two examples which further illustrate the nature of convergence of the absolutely expedient algorithms. In all the examples $E[\eta(k)|p(k)]$ is calculated as the numerical average of fifty sample experiments.

Example 3.4:   Let M = 2, D = (.8, .4). Two algorithms L^A_{R-I} and N^A_{R-P} are simulated, our choice of functions $\lambda(.)$ and $\mu(.)$ being the same as those of example 3.3.

Table 3.4. L^A_{R-I}: $\lambda(p) \equiv 0.1$ $\mu(p) \equiv 0$

k	$E[\eta(k) \mid p(k)]$
0	0.6000
25	0.6618
50	0.7203
75	0.7670
100	0.7800
125	0.7912
150	0.7934
175	0.7977
200	0.7993
225	0.7998
250	0.8000

Comparing Tables 3.4 and 3.5 we can conclude that with the above choice of functions the L^A_{R-I} and N^A_{R-P} algorithm have very nearly the same behavior as far as the variation of $E[\eta(k)|p(k)]$ with k. However, a quick look at the tables 3.2 and 3.3 reveals that the lower bound on $\Gamma_\ell(p)$ in the case of N^A_{R-P} algorithm is uniformly better than the corresponding lower bound in the case of L^A_{R-I} algorithm.

Table 3.5. N_{R-P}^{A}: $\lambda(p) = 0.05 + 0.05\ p_1 p_2$, $\mu(p) = 0.1\ p_1 p_2$

k	$E[\eta(k)\ |p(k)]$
0	0.6000
25	0.6663
50	0.7268
75	0.7675
100	0.7799
125	0.7911
150	0.7931
175	0.7965
200	0.7990
225	0.7997
250	0.7999

Example 3.5: M = 4; D = (0.90, 0.55, 0.40, 0.05) Variation of $E[\eta(k)|p(k)]$ as a function of k are given in table 3.6 and 3.7.

Table 3.6 L_{R-I}^{A}: $\lambda(p) \equiv 0.1$ $\mu(p) \equiv 0$

k	$E[\eta(k)\ |p(k)]$
0	0.4750
25	0.7797
50	0.8740
75	0.8987
100	0.8998
125	0.9000

Table 3.7 L_{R-I}^{A}: $\lambda(p) \equiv 0.01$, $\mu(p) \equiv 0$

K	[E $\eta(k)$	$p(k)$]
0	0.4750
100	0.5626
200	0.6468
300	0.7133
400	0.7628
500	0.7963
600	0.8277
700	0.8491
800	0.8620
900	0.8725
1000	0.8811
1100	0.8863
1200	0.8897
1300	0.8909
1400	0.8941
1500	0.8951

For the same D vector of this example, the following table 3.8 compares the lower bound on $\Gamma_{\ell}(p)$

Table 3.8

P_ℓ	Lower bound on $\Gamma_\ell(p)$	
	L_{R-I}^A Algorithm: $\lambda(p) = 0.1$ $\mu(p)$ 0	L_{R-I}^A Algorithm: $\lambda(p) = 0.01$ $\mu(p) = 0$
.1	0.5995	≈ 1.0
.2	0.8396	1.0
.3	0.9358	1.0
.4	0.9744	1.0
.5	0.9898	1.0
.6	0.9959	1.0
.7	0.9985	1.0
.8	≈ 1.0	1.0
.9	≈ 1.0	1.0

From table 3.6 and 3.7 it follows that the smaller the value of $\lambda(.)$, slower is the speed of convergence but higher is the lower bound on $\Gamma_\ell(p)$ as seen in table 3.8. This example clearly illustrates the trade-off between "speed" and "accuracy" as reflected by the lower bound on $\Gamma_\ell(p)$.

3.8 Comments and Historical Remarks:

Much of the material of this chapter 3 follows chapters 2, 3 and 5 author's Ph.D. Thesis [L4] and has appeared in condensed form in [L1] [L2] [L5].

Some examples of the algorithms satisfying our general symmetry conditions have earlier appeared in systems literature even before absolute expediency was known. Perhaps the earliest is due to Vorontsova [V8] who considered the case of $M = 2$ and $\lambda(p) = p_1 p_2$ and $\mu(p) = p_1 p_2$. Shapiro and Narendra [S7] independently discovered the L_{R-I}^{A} algorithm $\lambda(p) \equiv a < 1$ and $\mu(p) \equiv 0$.

While our approach to derive the bounds closely follows that of Norman [N11] [N12] a major point of departure is that our algorithm (3.1) or its variant (3.1b) are not necessarily distance diminishing. In [N11] Norman originally developed these ideas for $M = 2$, $\lambda(p) \equiv a < 1$ and $\mu(p) \equiv 0$, that is, in the context of L_{R-I}^{A} algorithms. Norman uses the distance diminishing property heavily to prove convergence with probability one. In our case symmetry conditions imply that the linear functional of the process is a sub-martingale and we exploit this latter property to derive convergence with probability one. Following Norman [N11] Baba and Sawaragi [B1] also independently demonstrated ε-optimality of two special cases of our general algorithm (3.1). Aso and Kimura [A3] have derived conditions for absolutely expediency for a general class of algorithm of the type (1.3) in chapter 1. (See exercise 3.12). So far absolutely expedient algorithms are the only sub class of the absorbing barrier algorithms that are well understood. Very recently Herkenrath, Kalin and Lakshmivarahan have obtained examples of absorbing barrier algorithms that are not absolutely expedient [H3].

3.9 APPENDIX

Completion of the proof of Theorem 3.1:

Necessary part:

To prove the necessity of (S3.1) - (S3.2) assume they are violated. That is, let

$$\frac{\eta_{ij}}{p_i p_j} = \frac{\phi_i}{p_i} - \frac{\phi_j}{p_j} \tag{3.A1}$$

$$\frac{\xi_{ij}}{p_i p_j} = \frac{\psi_i}{p_i} - \frac{\psi_j}{p_j} \tag{3.A2}$$

for all i, j = 1,2,...,M, where $\underline{\eta} = [\eta_{ij}]$ and $\underline{\xi} = [\xi_{ij}]$ are two M x M matrices of continuous otherwise arbitrary functions and $\eta_{ij} = \eta_{ij}(p)$ and $\xi_{ij} = \xi_{ij}(p)$ for simplicity. Evidently,

$$\frac{\eta_{tj}}{p_t p_j} = \frac{\eta_{ij}}{p_i p_j} - \frac{\eta_{it}}{p_i p_t} \tag{3.A3}$$

$$\frac{\xi_{tj}}{p_t p_j} = \frac{\xi_{ij}}{p_i p_j} - \frac{\xi_{it}}{p_i p_t} \tag{3.A4}$$

From (3.A3) - (3.A4) it follows that η_{ij} and ξ_{ij} for any given i and j = 1,2,...,M determine all the other η_{tj} and ξ_{tj} t = 1,2,...,M, j = 1,2,...,M. In other words, given any one row in each of the matrices $\underline{\eta}$ and $\underline{\xi}$, all the other remaining rows in $\underline{\eta}$ and $\underline{\xi}$ can be expressed in terms of that single row of $\underline{\eta}$ and $\underline{\xi}$ respectively. From this and from (3.A1) - (3.A2) it follows that $\underline{\eta}$ and $\underline{\xi}$ are skew-symmetric matrices of rank one.

Multiplying both sides of (3.A1) and (3.A2) by $p_i p_j$ and substituting in (3.6), the latter becomes

$$W_i(p) = p_i \sum_{i=1}^{M} [\phi_j + \psi_j] (d_i - d_j) + \sum_{J=1}^{M} B_{ij}$$

where $\qquad\qquad\qquad\qquad\qquad\qquad\qquad\qquad\qquad\qquad\qquad\qquad$ (3.A5)

$$B_{ij} = c_j \xi_{ij} - d_j \eta_{ij}$$

To prove the necessity, consider a vector $D_1 \in \mathcal{D}$ in which $d_i > d_j$, $j \neq i$ $j = 1,2,\ldots,M$, that is for this vector D_1 the value of $\ell = i$. Theorem 3.1 requires that for this vector, $W_i(p) > 0$. From (3.A5) it is seen that $W_i(p) > 0$ so long as

$$\sum_{j=1}^{M} B_{ij} > - \delta_i$$ (3.A6)

where $\delta_i = \delta_i[p,D_1] > 0$ and $\delta_i \to 0$ as $(d_i - d_j) \to 0$ for this given i.

Now consider another vector $D_2 \in \mathcal{D}$ in which all the components d_j $(j \neq i)$ have the same value as in vector D_1 but $d_i < d_j$ $j \neq i$. In other words for this vector m takes the value i. According to theorem 3.1 for this vector D_2, $W_i(p) < 0$. Again from (3.A5) it can be seen that $W_i(p) < 0$ so long as

$$\sum_{j=1}^{M} B_{ij} < \delta'_i$$ (3.A7)

where $\delta'_i = \delta_i'[p,D_2] > 0$ and $\delta'_i \to 0$ as $(d_j - d_i) \to 0$.

Notice all the components $d_j (j \neq i)$ are the same in both the vectors D_1 and D_2. From (3.A5) as B_{ij} depend only on d_j $(j \neq 1)$, it follows that (for the same p) $\sum_{j=1}^{M} B_{ij}$ is the same for both the vectors D_1 and D_2. Thus for the vectors D_1 and D_2 considered together sign definiteness of $W_\ell(p)$ and $W_m(p)$ follows only if

$$-\delta_i < \sum_{j=1}^{M} B_{ij} < \delta'_i$$ (3.A8)

As $(d_i - d_j)$ can take arbitrarily small values and as δ_i and $\delta'_i \to 0$ when $|d_i - d_j| \to 0$, the only way to satisfy the inequality (3.A8) is to choose

ε_{ij} and η_{ij} such that

$$\sum_{j=1}^{M} B_{ij} \equiv 0 \qquad (3.A9)$$

for this fixed value of i.

Now, if we consider the ensemble of all the vectors belonging to $\mathcal{D}$, it is easily seen that i can take all the values from $1,2,\ldots,M$. From this and from (3.A9) it follows that

$$\sum_{j=1}^{M} B_{ij} \equiv 0 \qquad \text{for all } i = 1,2,\ldots,M \qquad (3.A10)$$

Expressing all the elements of $\underline{\eta}$ and $\underline{\xi}$, without loss of generality, in terms their first row, using (3.A3) and (3.A4), it can be seen that (3.A10) implies

$$\eta_{1i}[1 - \sum_{j=1}^{M} p_j(1-d_j)] = \xi_{1j} \sum_{j=1}^{M} p_j(1-d_j) \qquad (3.A11)$$

for all $i = 1,2,\ldots,M$. Recall (3.A11) must be true for all $p \in S_M^o$ and $D \in \mathcal{D}$. The only choice of functions for which (3.A11) holds for all $p \in S_M^o$ and $D \in \mathcal{D}$ are

$$\eta_{1i} \equiv 0 \quad \xi_{1i} \equiv 0 \qquad \text{for all } i = 1,2,\ldots,M \qquad (3.A11)$$

This in turn in view of the properties of the matrix $\underline{\eta}$ and $\underline{\xi}$ implies that $\eta_{ij} = \xi_{ij} = 0$ for all $i,j = 1,2,\ldots,M$. Hence the necessity. Q.E.D.

Proof of Proposition 3.1:

Following section 5.D of Norman [N11] we shall prove this important proposition through a series of lemmas.

Lemma 3B.1 The Markov process generated by algorithm (3.1) converges to V_M with probability one.

Notice this is the same as theorem 3.4. The above lemma implies the condition (H.10) in Norman [N11]. An implication of this lemma is that

$$\lim_{n \to \infty} K^{(n)}[p,A] \text{ converges and}$$

$$\lim_{n \to \infty} K^{(n)}[p,A] \underset{=}{\Delta} K^{\infty}[p,A] \text{ is such that}$$

$K^{\infty}[p,V_M] = 1$ for all $p \in S_M$. In other words, $K^{\infty}[p,.]$ as a probability measure is concentrated only on V_M.

Lemma 3B.2: The operator u has no other eigenvalue of modulus 1 other than 1.

Proof: Let $uf = \bar{\lambda}f$ where $|\bar{\lambda}| = 1$ and $\bar{\lambda} \neq 1$, $f \in C[S_M]$. Let p_o be a point in S_M at which the function $f(.)$ attains a value whose absolute value is equal to the sup norm of $f(.)$ in S_M. That is

$$|f(p_o)| = \sup_{p \in S_M} |f(p)| \underset{=}{\Delta} |f|$$

Define

$$R_k = \{p \mid p \in S_M, \ f(p) = \bar{\lambda}^k f(p_o)\}$$
where $k = 1,2,3....$

As

$u^k f(p_o) = \bar{\lambda}^k f(p_o)$, we easily see that

$K^{(k)}[p_o, R_k] = 1$

Define

$$T_k(p) = \{p' \mid K^{(k)} [p, \{p'\}] > 0 \} \tag{B1}$$

In other words, $T_k(p)$ constitutes the set of all states that the Markov process visits with non-zero probability in k-steps starting from p.

Clearly, from the definition

$$T_k(p_o) \subset R_k$$

By lemma 3.B1 we know that there exists a sequence p(k) such that $p(k) \in T_k$ and $\lim_{k \to \infty} p(k) = e_j(p_o)$, a random absorbing state. Hence

$$\lim_{k \to \infty} f(p(k)) = f(e_{j(p_o)})$$

But $p(k) \in R_k$ and so $f(p(k)) = \bar{\lambda}^k f(p_o)$. This converges only if $f(p_o) = 0$ since $\bar{\lambda} \neq 1$ and $|\bar{\lambda}| = 1$. This in turn implies $f(p) \equiv 0$.

Hence $|\bar{\lambda}| = 1$ and $\bar{\lambda} \neq 1$ cannot be an eigenvalue of u.   Q.E.D.

<u>Lemma 3.B3</u>: <u>If b_1, b_2, ... b_M are M scalars there is one and only one</u>

$f(.) \in C[S_M]$ <u>such that</u>

$$uf = f$$

<u>and</u>

$$f(e_i) = b_i, \quad i = 1,2,\ldots,M$$

<u>Proof</u>:

<u>Uniqueness</u>: If $f(.) \in C [S_m]$ and uf = f, as u is an averaging operator all the maxima of $f(.)$ can occur only on V_M. Let p_o be the state such that

$$|f(p_o)| = \sup_{p \in S_M} |f(p)| \triangleq |f|$$

Let

$$R = \{p \mid f(p) = f(p_o)\}$$

Since $u^k f(p_o) = f(p_o)$, we have $K^{(k)} [p,R] = 1$

and $T_k(p_o) \subset R$ where $T_k(p)$ is defined in (B1) above. Again by lemma 3.B1 there is a sequence p(k) with p(o) = p_o such that $p(k) \in T_k(p_o)$ and

$$\lim_{k \to \infty} p(k) = e_{j(p_o)}, \text{ a random absorbing state.}$$

Hence

$$\lim_{k \to \infty} f(p(k)) = f(e_{j(p_o)})$$

But $p(k) \in R$ and so $f(p(k)) = f(p_o)$. Thus $f(p_o) = f(e_{j(p_o)})$ and $f(e_{j(p_o)}) = |f|$.

Now let g_1 and $g_2 \in {}^C[S_M]$ such that

$$u g_1 = g_1 \quad \text{and} \quad u g_2 = g_2$$

$$g_1(p) = g_2(p) \quad \text{for all } p \in V_M.$$

Let $g = g_1 - g_2$. Then $g \in C[S_M]$ and $g(p) = 0$ for all $p \in V_M$ and $ug = g$. From the above discussion it follows that $|g| = 0$ and so $g_1(p) = g_2(p)$ for all $p \in S_M$.

Existence:

Since $u^k f(p) = E[f(p(k+1)) \mid p(o) = p]$

we have

$$u^k f(p) = f(p) \quad \text{for all } p \in V_M$$

Thus

$$u_1 f(p) \triangleq \lim_{k \to \infty} u^k f(p) = f(p) \quad \text{for all } p \in V_M$$

Define functions

$$E_i(p) = [1 - \frac{|p-ei|}{\sqrt{2}}]^+$$

where $|x - y|$ is the Euclidian distance between x and y and

$$x^+ = \begin{cases} x & \text{if } x > 0 \\ 0 & \text{if } x \leq 0 \end{cases}$$

Let

$$h(p) = \sum_{i=1}^{M} b_i \, u_1 \, E_i(p)$$

Then

$$h(e_j) = \sum_{i=1}^{M} b_i \, u_1 \, E_i \, [ej] = bj$$

also

$$\ddot{u} h(p) = \sum_{i=1}^{M} b_i \, u \, u_1 \, E_i \, (p)$$

$$= \sum_{i=1}^{M} b_i \, u_1 \, E_i(p)$$

$$= h(p)$$

Hence $h(p)$ is the function that is sought.

The proposition 3.1 now easily follows from lemma 3.B3.　　　Q.E.D.

3.10 Exercises

3.1 Show that the classical linear-reward-penalty L_{R-P}^E algorithm and N_{R-P}^E algorithm discussed in chapter 2 are not absolutely expedient.

3.2 Complete the proof of theorem 3.3.

3.3 a. Show that the conclusions of the theorem 3.3 remains true even if we extend the conditions as $D \varepsilon \mathcal{D}_2$ where

$\mathcal{D}_2 = \{D | D = (d_1, d_2, \ldots, d_M), \ 0 \le d_i < 1, \ d_i \ne d_j$

In other words d_m is allowed to be 0.

b. Investigate the effect of the conditions $(C'3.3)$ not being true on the conclusions of theorem 3.3.

c. What happens when some of the components of D are equal?

3.4 Show that the algorithm (3.1) is in general not distance diminishing. If $\lambda(.) \equiv a < 1$ and $\mu(.) \equiv 0$ then we get L_{R-I}^A algorithm. Verify that this algorithm is distance diminishing.

3.5 Verify the properties of the functions $v[u]$ and $H[u]$ and $\dfrac{d\ H[u]}{du}$ and plot these functions.

3.6 From the properties of the function $H[u]$ and from (3.49) derive the inequalities in (3.50).

Hint: Recall if $a > 0$ and $b > 0$ and denoting $H'[u] = \dfrac{d\ H[u]}{du}$ we get $H'[-(a+b)] < H'[-b] < H'[0] < H'[a] < H'[(a+b)]$.

3.7 Show that $\dfrac{d\ H[u]}{du} < 1 - \dfrac{1}{2} e^{-0.2u}$ for all $u > 0$

Hint: Let $\dfrac{d\ H[u]}{du} < 1 - \dfrac{1}{2} e^{-\alpha u}$ for some $\alpha > 0$.

Rewriting this inequality as

$$\alpha > \frac{1}{u} \log_e \left[\frac{u \, [e^u - 1]}{2[u - e^u + 1]} \right]$$

Verify by direct computation that the supremum of the R.H.S. of the last inequality for all $u > 0$ is ≈ 0.193.

3.8 If $p(k) = p$, then the conditional variance of $p_i(k+1)$ denoted by $\sigma_i^2(p)$ is given by

$$\sigma_i^2(p) = E[p_i^2(k + 1) \mid p(k) = p] - E^2[p_i(k + 1) \mid p(k) = p]$$

Show that

$$\sigma_i^2(p) = p_i(1-p_i) \, \lambda^2(p)$$

$$+ \, [\lambda^2(p) - \mu^2(p)] \, [p_i(1-p_i)^2 \, c_i + p_i^2 \sum_{j \neq 1} p_j \, c_j]$$

$$- \, [\lambda^2(p) + \mu^2(p)] \sum_{j \neq 1} p_i \, p_j \, (d_i - d_j)^2$$

Hence or otherwise show that $\sigma_i^2(p) = 0$ if and only if $p \, \epsilon \, V_M$. Note this is another way to show that V_M forms the only absorbing states of the algorithm (3.1B). Verify that each member of V_M is an ergodic kernel.

3.9 a. Show that the existence of upper bound on $\Gamma_\ell(p)$ of the form (3.36) implies that the N_{R-P}^A algorithm is not optimal.

b. Plot the function $\dfrac{1-e^{-xp}}{1-e^{-x}}$ as a function of $p \epsilon \, [0,1]$ for various values of x say $x = 0.1, 0.5, 1.0, 5.0, 10.0, 25.0$.

3.10 Show that

$$\delta\eta(k) = \frac{1}{2} [\lambda(p) + \mu(p) \; p^T A \; p$$

where $A = [a_{ij}] = [(d_j - d_i)^2]$

Investigate the properties of the quadratic form $p^T A p$ for all $p \, \varepsilon \, S_M$.

3.11 If $\lambda(p) \equiv 0$ and $\mu(p) \not\equiv 0$ for all $p \, \varepsilon \, S_M$, the resulting algorithm is called non-linear inaction-penalty algorithm.

a) Show that every point on the boundary of the simplex S_M is an absorbing state for this algorithm. (Hence this is not an absorbing barrier algorithm in the sense of the definition).

b) Hence or otherwise show that for the above choice of $\lambda[.]$ and $\mu[.]$

$$E[\eta(k+1) \mid p(k)] \geq \eta(k)$$ with probability one with equality holding good for all $p(k)$ belonging to the boundary of the simplex. (Compare this with the definition of absolute expediency given in chapter 1).

c) Compare and catalogue all the properties of N^A_{R-P}, N^A_{R-I} and the non-linear inaction-penalty algorithm introduced in this exercise.

3.12 Consider the following generalization of the algorithm (3.1):

$$p_i(k+1) = p_i(k) + \sum_{j \neq i} \phi_j \; [i, p(k)]$$
$$p_j(k+1) = p_j(k) - \phi_j \; [i, p(k)]$$
$(j \neq i)$

if coin i is chosen and toss resulted in success.

$$p_i(k+1) = p_i(k) - \sum_{j \neq i} \psi_j [i, p(k)]$$
$$p_j(k+1) = p_j(k) + \psi_j \; [i, p(k)]$$
$(j \neq i)$

if coin i is chosen and toss resulted in failure.

where $\phi_s: I \times S_M \to [0,1]$ and $\psi_s: I \times S_M \to [0,1]$

where $I = \{1,2,\ldots,M\}$.

Find a set of necessary and conditions for absolute expediency for the above general algorithm. (Aso and Kimura [A3]).

3.13 (a)   Find a set of necessary and sufficient conditions for the general

algorithm presented in exercise (3.12) to be an absorbing barrier

algorithm. (Herkenrath, Kalin, Lakshmivarahan [H3]).

(b)   Hence or otherwise find a set of necessary and sufficient condi-

tions for the algorithm (3.1) to be an absorbing barrier algorithm

and compare such conditions with the symmetry conditions (S3.1)

and (S3.2) of theorem 3.1.

3.14   Q-Model Learning Algorithm [L3].   Consider a family M independent

populations with unknown distribution.   If x_i a random variable from

the ith population, then

$$\left. \begin{array}{l} \text{Prob } [x_i = 1(\text{win})] \triangleq d_i \\[6pt] \text{Prob } [x_i = 0 \ (\text{draw})] \triangleq 1-c_i-d_i \\[6pt] \text{Prob } [x_i = -1 \ (\text{loss})] = c_i \end{array} \right\} \quad \forall_i = 1,2,\ldots,M$$

Let $p(k) = (p_1(k), p_2(k), \ldots, p_M(k)^T$ where $p(k) \, \varepsilon \, S_M$ and $p_i(k)$

refers to the probability that at time k, the ith population is

sampled.   Define

$$\eta(k) = \sum_{i=1}^{M} p_i(k) \ E[x_i]$$

p(k) is updated according to the following algorithm:

If the ith population is sampled at time instant k, then

$$\left. \begin{array}{l} p_i(k+1) = p_i(k) + \sum_{j \neq i} \phi_j \ [p(k)] \\[10pt] p_j(k+1) = p_j(k) - \phi_j \ [p(k)], \ j \neq i \end{array} \right\} \quad \text{if } x_i = 1$$

$$\left. \begin{array}{l} p_i(k+1) = p_i(k) + \sum_{j \neq i} \chi_i [p(k)] \\[10pt] p_j(k+1) = p_j(k) - \chi_j [p(k)], j \neq i \end{array} \right\} \quad \text{if } x_i = 0$$

$$p_i(k+1) = p_i(k) - \sum_{j \neq i} \psi_j[p(k)] \left.\begin{matrix}\\ \\\end{matrix}\right\} \quad \text{if } x_i = -1$$

$$p_j(k+1) = p_j(k) + \psi_j[p(k)], j \neq i$$

(a) Find a set of necessary and conditions for absolute expediency.

(b) Drive the bounds on $\Gamma_i(p)$ for the above Q-model algorithm where $\Gamma_i(p)$ is defined in (3.22).

3.15 <u>S-Model Learning Algorithm</u> [L 3] If the M independent populations introduced in the previous exercise are such that the random variables from each population have a continuous distribution in [0,1] with <u>unknown</u> mean, we obtain the so-called S-model. Thus, if x_i is a random variable from the ith population, then let

E $[x_i] = s_i$ and s_i is unknown where $i = 1,2,...,M$

a) Design a learning algorithm of the absorbing barrier class to suit the S-model description given above and find necessary and sufficient conditions for absolute expediency where

$$\eta(k) = \sum_{i=1} p_i(k) \, E[x_i].$$

b) Find conditions for ε-optimality for your S-model algorithm.

3.16 In the main body of chapter 3 motivated by coin tossing, we considered the case when d_i's are all constants. If d_i's are time-varying and

$$d_1(k) + \delta > d_i(k) \quad \text{for all } i \neq 1 \, , \, \delta > 0$$

then using the L_{R-I}^A algorithm, $(\lambda(p) \equiv a, \mu(p) \equiv 0$ and $0 < a < 1)$ show that for every $\varepsilon > 0$ there exists an a* < 1 such that for all a < a*

$$\lim_{k \to \infty} | E [p_1(k)] - 1 | < \varepsilon.$$

(Refer Baba and Sawaragi [B1]).

CHAPTER 4

TIME VARYING LEARNING ALGORITHMS

4.1 Introduction:

In this chapter we analyze the time varying analogues of the Ergodic
learning algorithms presented in Chapter 2. These algorithms are obtained
from algorithm (2.1) by letting the step length parameter θ to vary with
time. The time-varying algorithms generate non-stationary Markov processes
over the simplex S_M. Our aim is to present different methods of asymptotic
analysis of these time varying learning algorithms. In the interest of
brevity we will only consider the case of $M = 2$, leaving the obvious extensions as exercises.

A general class of time-varying algorithm is described in section 4.2
where using the now classical methods of stochastic approximations it is
shown that the Markov process $\{P(k)\}$ converges with probability one. As a
by product it follows from this result that $\lim_{k \to \infty} \eta(k) = \eta^*$ and η^* can be made
as close to $\max_j \{d_j\}$ as desired. In Section 4.3 we present an alternate method
which combines some of the results of Chapter 2 and section 4.2. In this
section our emphasis is to bring out the basic ideas of the powerful techniques which were recently developed by Kushner [K11] and described in
detail in Kushner and Clark [K10]. As a first step the sequence $P(k)$ is

linearly interpolated into a continuous parameter process in $[0,\infty)$, with
the interpolation intervals being given by θ_k, the step-length parameter
at time k. Then we define a sequence of left shifts of the interpolated
process, the amount of the shift is given by $\{t_k\}$ where $t_k = \sum\limits_{i=0}^{k-1} \theta_i$.
Letting $\theta_k \to 0$ as $k \to \infty$ we obtain increasing compressions of discrete time
to continuous time as $k \to \infty$. It is shown that increasing compressions tends
to eliminate the noise effect in the long run. Further by requiring $t_k \to \infty$
as $k \to \infty$, the left-shifts bring the asymptotic part of the interpolated
continuous time process in the neighborhood of the origin. Using (a) Arzela-
Ascoli theorem or (b) weak convergence of probability measures associated
with the sequence of left shifts, a convergent subsequence of leftshifts of
interpolated process is obtained. It follows that the limit of these convergent subsequences are easily characterized by an ordinary differential
equation. The asymptotic properties of $\{P(k)\}$ are intimately related to the
asymptotically stable solutions of the above ordinary differential equations.
This method is known as compactness method as it uses either the Arzela-
Ascoli theorem [D3] or "relative Compactness" of a sequence of probability
measures [B5] [P1].

A brief discussion of the weak convergence ideas are presented in
Appendix 4A.

4.2. A time Varying Learning Algorithm:

As in Chapter 2 (section 2.3) let $p = (P, 1-P)^T$ and let[1]

$$
P(k+1) - P(k) = \begin{cases} \theta_k \; \lambda[P(k)] \; [1-P(k)] & \text{if coin 1 tossed and success} \\ -\theta_k \; \psi_1[P(k)] & \text{if coin 1 tossed and failure} \\ \theta_k \; \psi_2[P(k)] & \text{if coin 2 tossed and failure} \\ -\theta_k \; \lambda[P(k)] \; P(k) & \text{if coin 2 tossed and success} \end{cases}
\tag{4.1}
$$

where

(C4.1) $\lambda(P)$ and $\psi_j(P)$ satisfy the conditions (C2.1) - (C2.3) (refer section

2.2, Chapter 2)

(C4.2) $(1-P) \psi_2(P) \gtreqless P\psi_1(P)$ for $P \lesseqgtr \frac{1}{2}$

(C4.3) $\theta_k \to 0, \sum\limits_{k=1}^{\infty} \theta_k = \infty, \; \sum\limits_{k=1}^{\infty} \theta_k^2 < \infty$

Further θ_k are chosen in such a way that $P(k+1) \; \epsilon \; [0,1]$ if $P(k) \; \epsilon \; [0,1]$.

Let $P(k+1) - P(k) \triangleq \theta_k \; \delta P(k)$ and let

$$
E[\delta P(k)] \mid P(k)] = W(P)
\tag{4.2}
$$

It can be seen from (4.1) that

$$
W(P) = W^R(P) + W^P(P)
$$

where

$$
W^R(P) = \lambda[P] \; P(1-P) \; (d_1 - d_2)
$$

$$
W^P(P) = [(1-P) \; \psi_2 \; (P) \; C_2 - P \; \psi_1(P) \; C_1]
$$
$$\tag{4.3}$$

If $\lambda(.)$ and $\psi_j(.)$ $j = 1,2$, satisfy the conditions (C4.1) - (C4.2), from

theorem 2.1 and corollary 2.2 it follows that

1) there exists a unique $\beta_2 \; \epsilon(0,1)$ such that

$$
(P - \beta_2) \; W(P) \leq 0
$$

where

$$
\beta_2 \gtreqless \frac{1}{2} \quad \text{according as } c_1 \lesseqgtr c_2
$$
$$\tag{4.4}$$

[1] Our notations follow those in Chapter 2.

and

2)

$$\frac{d\ W(\beta_2)}{dp} < 0 \tag{4.5}$$

With these preliminary observations we shall now state and prove the following:

Theorem 4.1: Under conditions (C4.1)-(C4.3) the non-stationary Markov process $\{P(k)\}_{k \geq 1}$ converges to β_2 with probability one, that is,

$$\lim_{k \to \infty} P(k) = \beta_2 \text{ with probability one.}$$

Proof: From (4.1) it can be seen that

$$E[\ (P(k+1) - \beta_2)^2 \mid P(k)\ ]$$
$$= (P(k) - \beta_2)^2 + 2\ \theta_k[P(k) - \beta_2]\ W\ [P(k)] + \theta_k^2\ E[(\delta P(k))^2 | P(k)] \tag{4.6}$$

Since

$(P(k) - \beta_2)\ W[P(k)] \leq 0$, the following inequality follows from (4.6):

$$E[(P(k+1)-\beta_2)^2 \mid P(k)] \leq (P(k) - \beta_2)^2 - 2\ \theta_k | (P(k)-\beta_2)\ W[P(k)]\ |$$
$$+ \theta_k^2\ E[(\delta P(k))^2 \mid P(k)] \tag{4.7}$$

Recall that $\lambda(.)$ and $\psi_j(.)$, $j = 1,2$ are continuous non-negative functions with compact support and hence are bounded. This in turn implies that both $W(P)$ and $E[\delta P(k))^2 \mid P(k)]$ are bounded. Using this fact the above inequality may be rewritten (after dropping the second term on the right hand side) as

$$E[(P(k+1) - \beta_2)^2 \mid P(k)] \leq (P(k)-\beta_2)^2 + L\ \theta_k^2 \tag{4.8}$$

where L is a suitable positive constant. Define

$$Y_k \triangleq (P(k)-\beta_2)^2 + L \sum_{j=k}^{\infty} \theta_j^2 \tag{4.9}$$

From (4.7) and (C4.3) it follows that Y_k is a sequence of non-negative bounded random variables. Combining (4.8) and (4.9) we get

$$E[Y_{k+1} \mid P(k)] \leq Y_k \tag{4.10}$$

From (4.9) we have

$$F[P(k), P(k-1), \ldots P(1)] \supset F[Y_k, Y_{k-1}, \ldots Y_1] \tag{4.11}$$

where $F[x_j, x_{j-1}, \ldots x_i]$ $j \geq i$ is the σ-field generated by the random variables listed within the parenthesis. Taking conditional expectations of both sides of (4.10) using (4.11) we obtain

$$E[Y_{k+1} \mid Y_k, Y_{k-1}, \ldots Y_1] \leq Y_k \tag{4.12}$$

Thus $\{Y_k\}$ is a non-negative, bounded super martingale and

$$0 \leq \ldots \ldots \leq E[Y_{k+1}] \leq E[Y_k] \leq \ldots \ldots \leq E[Y_1] < \infty$$

By super martingale theorem, the sequence Y_k converges to a random variable with probability one and hence from (4.9), $(P(k)-\beta_2)^2$ must converge to a random variable, say ξ, with probability one. [D2].

Now it only remains to show that Prob $[\xi = 0] = 1$. To this end take expectations on both sides of (4.6):

$$E[(P(k+1) - \beta_2)^2] = E[(P(k) - \beta_2)^2]$$
$$- 2\theta_k E[\mid (P(k) - \beta_2) W[P(k))] \mid]$$
$$+ \theta_k^2 E[(\delta P(k))^2] \tag{4.13}$$
$$\leq (P(0) - \beta_2)^2 - 2 \sum_{j=0}^{k} \theta_j E[\mid (P(j) - \beta_2) W[P(j)]\mid] + L \sum_{j=0}^{k} \theta_k^2 \tag{4.14}$$

where P(0) is a constant with probability one. From (4.14) and (C4.3) it follows that

$$\sum_{j=0}^{\infty} \theta_j E[\mid (P(j) - \beta_2) W[P(j)] \mid] < \infty \tag{4.15}$$

since $E[(P(k) - \beta_2)^2] < \infty$ for all k. As

$$\sum_{j=0}^{\infty} \theta_j = \infty \text{ and } \mid P(j) - \beta_2 \mid W[P(j)] \geq 0 \text{ for all } j,$$

from (4.15) and exercise (4.1), it follows that there exists a subsequence k_i such that

$$\lim_{i \to \infty} E[\ |(P(k_i) - \beta_2) \ W[P(k_i)]| \ ] = 0 \qquad (4.16)$$

From exercise (4.2) it follows that

$$\lim_{i \to \infty} | \ (P(k_i)-\beta_2) \ W[P(k_i)] \ | = 0 \quad \text{in probability.} \qquad (4.17)$$

From exercise (4.3) there exists a further subsequence (say n_i) such that

$$\lim_{i \to \infty} | \ (P \ (n_i) - \beta_2) \ W[P(n_i)| = 0 \quad \text{with probability 1} \qquad (4.18)$$

From (4.4) and (4.18) we see that $\lim_{i \to \infty} P(n_i) = \beta_2$

with probability one. Since P(k) converges with probability one and as every converging subsequence must converge to the same limit as P(k) we obtain $\lim_{k \to \infty} P(k) = \beta_2$ with probability one. (Q.E.D.)

<u>Remark 4.1</u>: An immediate and important consequence of the above theorem is that

$$n^* \triangleq \lim_{k \to \infty} \bar{n}(k) = \lim_{k \to \infty}[P(k) \ d_1 + (1-P(k) \ ) \ d_2]$$

$$= \beta_2 \ d_1 + (1 - \beta_2) \ d_2 \quad \text{with probability one.}$$

$$> \frac{d_1 + d_2}{2} \quad \text{with probability one.} \qquad (4.19)$$

where $\beta_2 \gtrless \frac{1}{2}$ when $d_1 \gtrless d_2$ (see 4.4 above). Further, by replacing $\psi_j(P)$ by $\alpha\psi_j(P)$ where α is real, in view of corollary (2.1), by choosing α small enough we can make β_2 as close to 1 or 0 depending on $d_1 > d_2$ or $d_1 < d_2$. In other words for every $\varepsilon > 0$, there exists an α^* such that for all $\alpha \leq \alpha^*$

$$|n^* - \max \ (d_1,d_2) \ | < \varepsilon \ \text{with probability one} \qquad (4.20)$$

A comparison of (4.19) and (4.20) with (1.7) and (1.9) motivates the following definitions:

Definition 4.1:   A learning algorithm is said to be

(a)   Strongly expedient if

$$\lim_{k \to \infty} \eta(k) > \frac{1}{M} \sum_{i=1}^{M} d_i \qquad (4.21)$$

and

(b)   Strongly ε-optimal if for every ε > 0

$$\lim_{k \to \infty} |\eta(k) - \max_{j} \{d_j\}| < \varepsilon \qquad (4.22)$$

The corollary 4.1 summarizes the above observations:

Corollary 4.1:   There exists proper choice of functions satisfying (C4.1) - (C4.3) such that the algorithm (4.1) is strongly ε-optimal.

4.3 Kushner's Method of Asymptotic Analysis:

In this section we apply a technique recently developed by Kushner [K11] and discussed in detail in Kushner and Clark [K10] in the context of general stochastic approximation methods, for the analysis of time varying learning algorithm (4.1).

As a first step let us rewrite the algorithm (4.1) as:

$$P(k+1) = P(k) + \theta_k \, W[P(k)] - \theta_k \, \xi(k)$$

where

$$\xi(k) = [W \, [P(k)] - \delta P(k)] \qquad (4.23)$$

Define a piece-wise linear interpolation $P^0(t)$ of $P(k)$ as

$$P^0(t_k) = P(k)$$
$$P^0(t) = \frac{(t_{k+1} - t)}{\theta_k} P(k) + \frac{(t-t_k)}{\theta_k} P(k+1) \qquad (4.24)$$

for $t \, \varepsilon \, (t_k, t_{k+1})$ where

$$t_k = \sum_{i=0}^{k=1} \theta_i \qquad (4.25)$$

Also define a piece-wise constant, right-continuous interpolation $\overline{P}^0(t)$ of

P(k) where

$$\overline{P}^0(t) = P(k) \text{ for } t \in [t_k, t_{k+1})$$

(4.26)

Similarly let

$$M^0(t_k) = \sum_{i=0}^{k-1} \theta_i \, \xi(i)$$

$$M^0(t) = \frac{(t_{k+1})-t}{\theta_k} \, M^0(t_k) + \frac{(t-t_k)}{\theta_k} \, M^0(t_{k+1})$$

(4.27)

for $t \in (t_k, t_{k+1})$

Now iterating (4.23) we obtain

$$P(k) = P(0) + \sum_{i=0}^{k-1} \theta_i \, W[P(i)] - \sum_{i=0}^{k-1} \theta_i \, \xi(i)$$

(4.28)

Using the interpolations we may rewrite (4.28) as

$$P^0(t) = P^0(0) + \int_0^t W[\overline{P}^0(s)] \, ds + M^0(t)$$

(4.29)

Let

$$e(t) \triangleq \left| \int_0^t W[\overline{P}^0(s)] \, ds - \sum_{i=0}^{k-1} \theta_i \, W[P(i)] \right|$$

(4.30)

where $e(t)$ evidently depends on $\{\theta_i\}$.

Since $W(.)$ is bounded and continuous (in view of C4.1), it can be seen

[A 1] by letting $\sup_{i \geq 0} \{\theta_i\} \to 0$ (in addition to C4.3), $k \to \infty$ and

$t = \sum_{i=0}^{k-1} \theta_i$ that $e(t) \to 0$ uniformly for all k in finite intervals.

Obtain a sequence of functions $P^k(t)$, $M^k(t)$ for $k = 0, 1, 2, \ldots$,

as left-shifts of $P^0(t)$ and $M^0(t)$ by

$$P^k(t) = \left\{ \begin{array}{l} P^0(t+t_k) \text{ for } t \geq t_k \\ P(0) \text{ for } t \leq - t_k \end{array} \right\}$$

(4.31)

$$M^k(t) = \left\{ \begin{array}{ll} M^0(t+t_k) - M^0(t_k) & \text{ for } t \geq t_k \\ - M^0(t_k) & \text{ for } t \leq -t_k \end{array} \right\}$$

(4.32)

Notice $P^k(0) = P^0(t_k) = P(k)$.

From (4.30) – (4.32) we get

$$P^k(t) = P^k(0) + \int_0^t W[\bar{P}^0(t_{k+s})]\ ds + M^k(t) \quad \text{for } t \geq t_k$$

$$= P(0) \quad \text{for } t \leq - t_k \tag{4.33}$$

Define

$$\int_0^t W[P^k(s)]\ ds = \int_0^t W[\bar{P}^0(t_{k+s})]\ ds + \varepsilon^k(t) \tag{4.34}$$

where $\varepsilon^k(t)$ is the error in replacing the piece-wise constant interpolation $\bar{P}^0(t)$ by the piece-wise linear interpolations $P^0(t)$. Again, as $W(.)$ is bounded and continuous we easily see [A6] that

$$\underset{t}{\text{Sup}}\ |\ \varepsilon^k(t)\ | \to 0 \quad \text{as} \quad \left\{ \begin{array}{l} k \to \infty \\ \\ \underset{i}{\sup}\ \{\theta_i\} \to 0 \end{array} \right. \tag{4.35}$$

(uniformly) for all t in finite intervals. Having observed this, we can rewrite (4.33) as

$$P^k(t) = P^k(0) + \int_0^t W[P^k(s)]\ ds + M^k(t); \quad t \geq t_k$$

$$= P(0) \qquad\qquad\qquad t \leq t_k \tag{4.36}$$

For later use we shall collect all the properties of $\xi(k)$ and $\sum_{i=0}^{k-1} a_i \xi(i)$

(P4.1) $E[\xi(k)\ |\ P(k)] = 0$ [follows from the definition of $W(.)$]

(P4.2) As $W(.)$ is a bounded function, the variance of $\xi(k)$ is uniformly bounded.

(P4.3) Since $P(k)$ is a Markov process and θ_k is a real sequence $\sum_{i=0}^{k} \theta_i \xi(i)$ is a martingale sequence. Hence by Martingale convergence theorem $\sum\limits_{i=0}^{k} \theta_i \, \xi(i)$ converges with probability one.

In summary, using the left-shifts of the interpolations of P(k) and $M^0(t_k)$, in this section we have obtained a continuous time version of the time varying learning algorithm (4.1). The conditions

(C4.3') $\theta_k \to 0$ as $k \to \infty$ and $\sum\limits_{i=0}^{\infty} \theta_i = \infty$

are very crucial for the derivation of (4.36). In addition to (C4.3') if the sequence θ_i is also square summable, that is, $\sum\limits_{i=0}^{\infty} \theta_i^2 < \infty$, it will be shown below that $M^0(t)$ is uniformly continuous on $[0,\infty)$ with probability one. Using compactness arguments (Arzela-Ascoli theorem) it will then be shown that $\lim\limits_{k \to \infty} P(k)$ converges (to β_2, refer (4.4) and (4.5)) with probability one. Without square summability we can only obtain a weaker result, namely $\lim\limits_{k \to \infty} P(k)$ converges (to β_2) in probability. However in either case it will be shown that the limit process (in the sense of Arzela-Ascoli theorem or weak convergence of measures induced by the left shifts of interpolations) satisfies the same ordinary differential equation (also same as equation 2.13).

A. Convergence with probability one:

In this sub section it will be assumed that conditions (C4.1)-(C4.3) hold. The next two lemmas characterize the properties of the continuous time process $M^0(t)$ resulting from these conditions.

Lemma 4.1   If θ_i satisfies the condition (C4.3) then

$$\lim_{k \to \infty} P[\sup_{m \geq k} \left| \sum_{i=k}^{m} \theta_i \, \xi(i) \right| \geq \epsilon] = 0 \tag{4.37}$$

Proof: If $\{x(m)\}$ is a Martingale sequence, then by Doob's inequality [D2] for any $\epsilon > 0$

$$P[\sup_{m \geq 0} \left| x(m) \right| \geq \epsilon] \leq \lim_{m \to \infty} \frac{E[x^2(m)]}{\epsilon^2}$$

Applying this to our process $\sum\limits_{i=0}^{k} \theta_i \, \xi(i)$, we obtain

$$P[\sup_{m \geq k} | \sum_{i=k}^{m} a_i \, \xi(i) | \geq \epsilon] \leq \lim_{m \to \infty} E[\sum_{i=k}^{m} \theta_i \, \xi(i)]^2 \frac{}{\epsilon}$$

$$\leq \frac{L \sum\limits_{i=k}^{\infty} \theta_i^2}{\epsilon} \qquad (4.38)$$

The second inequality follows from the property (P4.2) listed above.

From (C4.3), as $k \to \infty$, the r. h. s. of (4.38) $\to 0$ from which the lemma

follows. Q.E.D.

Combining (4.27) and (4.37) we get

$$\lim_{k \to 0} P [\sup_{m \geq k} | M^0(t_m) - M^0(t_k) | \geq \epsilon] = 0 \qquad (4.39)$$

Lemma 4.2: (4.39) implies that $M^0(.)$ is uniformly continuous with proba-

bility one on $[o, \infty)$.

Proof: By hypothesis for all $n > 0$

$$\lim_{k \to \infty} P [\sup_{m \geq k} | M^0(t_m) - M^0(t_k) | \geq \frac{1}{2^n}] = 0$$

It follows that for each n we can find k_n such that

$$P[\sup_{m \geq k_n} | M^0(t_m) - M^0(t_{k_n})| \geq \frac{1}{2^n}] \leq \frac{1}{2^n}$$

and consequently

$$\sum_n P[\sup_{m \geq k_n} | M^0(t_m) - M^0(t_{k_n}) | \geq \frac{1}{2^n}]$$

$$\leq \sum_n \frac{1}{2^n} < \infty$$

Thus, by Borel-Cantelli lemma [T7]

$$\sup_{\substack{i > j \\ j \geq k_n}} | M^0(t_i) - M^0(t_j)| < \frac{1}{2^n} \qquad (4.40)$$

for all but finite number of n with probability one. This together with the
continuity with probability one of $M^0(.)$ on $[0,\infty)$ imply that for every $t > 0$

$$\sup_{\substack{|t-s|<T \\ s,t \geq k_n}} \quad | M^0(t) - M^0(s)| < \frac{1}{2^n}$$

(4.41)

for all but finite number of n with probability one. Hence the lemma 4.2.

Q.E.D.

With these preliminaries we can state the basic convergence theorem.

Theorem 4.2: Let P(k) be defined by (4.1) and assume the conditions (C4.1)
- (C4.3). Then

(a) With probability one $\{P^k(t)\}$ is equicontinuous and that the
limit $\{P(t)\}$ of any convergent subsequence of $\{P^k(t)\}$ satisfies the
ordinary differential equation

$$\frac{dP(t)}{dt} = W[P(t)]$$

(4.42)

in the time interval $(-\infty, \infty)$.

(b) $\lim_{k \to \infty} P(k) \; \epsilon \; S$ with probability one where S consists of the set of
all asymptotically stable solutions of (4.42).

Proof: We have already seen that

$$P^k(t) = P^k(0) + \int_0^t W[P^k(s)] \, ds + M^k(t); \; t \geq t_k$$

$$= P(0) \qquad\qquad\qquad t \leq t_k$$

Recall

1) $P^0(t)$ is bounded on $[0,\infty)$ and hence $\{P^k(t)\}$ is uniformly bounded
on $(-\infty, \infty)$.

2) From lemma 4.2, $M^0(.)$ is uniformly continuous and hence $M^k(.)$ are
uniformly continuous on $(-\infty, \infty)$, bounded on finite intervals in
$(-\infty, \infty)$ and $M^k(.) \to 0$ as $k \to \infty$ uniformly on finite intervals in

$(-\infty, \infty)$

Thus $\{P^k(t)\}$ is uniformly bounded and equicontinuous (uniform continuity of $P^k(.)$ in $(-\infty, \infty)$ for each k) on $(-\infty, \infty)$. By Arzela-Ascoli theorem [D3] we can extract a convergent subsequence indexed by k_i such that

$$\lim_{i \to \infty} P^{k_i}(t) = P(t) \quad \text{exists.}$$

Hence $P(t)$ satisfies the equation

$$P(t) = P(0) + \int_0^t W[P(s)] \, ds$$

which is exactly (4.42)

To prove (b), notice that under conditions (C4.1) and (C4.2), β_2 is the only asymptotically stable solution of (4.42) (refer (4.4) and (4.5)). Thus by asymptotic stability theorem [L11]

$$\lim_{t \to \infty} P(t) = \beta_2 = \lim_{k \to \infty} P(k), \text{ with probability 1.} \qquad \text{Q.E.D.}$$

Remark 4.1: It should be interesting to note that the differential equation (2.13) and (4.42) are one and the same equation. In theorem 2.2, it was shown that the solution of this differential equation well approximates the evolution of the mean of the process $\{P(k)\}$. But in theorem 4.2 it is shown that the sample path of the process $\{P(k)\}$ after suitable transformations (interpolation and left shifting) in fact coincides with the solution of the same differential equation, with probability one. Obviously, conclusions of the theorem 4.2 are much stronger but the trade-off is that the algorithm (4.1) is a time varying algorithm. Also notice that the square summability of θ_i's is very crucial for the above conclusion.

B. Weak Convergence:

In this subsection we investigate the effect of non-square summability of θ_i's. Assume (C4.1), (C4.2) and (C4.3'). Our next lemma examines the properties of $M^0(t)$ resulting from the condition (C4.3').

Lemma 4.3: Assume (C4.3'). Then,

$$\lim_{k \to \infty} P[\max_{k \le m \le N} | \sum_{i=k}^{m} \theta_i \xi(i) | \ge \epsilon] = 0 \tag{4.43}$$

Proof:

From properties (P4.1)-(P4.3) we have for each k, the sequence

$\{ \sum_{i=k}^{m} a_i \xi(i), m \ge k\}$ is a martingale sequence. Further from (P4.2)

$$E[| \xi(k+1)|^2 | \xi_i, 0 \le i \le k] \le \sigma^2 \text{ with probability 1} \tag{4.44}$$

By Doob's inequality, for every $\epsilon > 0$

$$P[\max_{k \le m \le N} | \sum_{i=k}^{m} \theta_i \xi(i)| \ge \epsilon] \le \frac{E| \sum_{i=n}^{N} \theta_i \xi(i)|^2}{\epsilon^2} \tag{4.45}$$

$$\le \left(\frac{\sigma^2}{\epsilon^2} \right) \left(\sum_{i=k}^{N} \theta_i \right) \sup_{i \ge k} \{\theta_i\}$$

For every $\infty > N \ge k$, the r.h.s. of this inequality is bounded and from (C.43') it follows that $\sup_{i \ge k} \{\theta_i\} \to 0$ as $k \to \infty$. Hence the lemma.   Q.E.D.

Define

$$m(t) = \begin{cases} \max \{n: t_n \le t\} & t \ge 0 \\ 0 & t \le 0 \end{cases} \tag{4.46}$$

Then, clearly

$$t_n + s \varepsilon \left[\sum_{i=0}^{-m(t_n+s)-1} \theta_i, \sum_{i=0}^{m(t_n+s)} \theta_i \right] \text{for } s > 0$$

and for any $0 < T < \infty$

$$\lim_{n \to \infty} \sum_{m(t_n)}^{m(t_n+s)-1} \theta_i = T$$

(4.47)

With these notations, from lemma 4.3 it follows that

$$\lim_{k \to \infty} P \left[\max_{0 \le t \le T} \left| \sum_{i=m(t_k)}^{m(t_k+t)} \theta_i \, \xi(i) \right| \ge \varepsilon \right] = 0$$

(4.48)

The following theorem 4.3 summarizes the properties of $P^0(t)$ resulting

from the non-summability of the sequence of θ_i's. It is typical of the

results and theorems concerning the weak convergence of probability measures

over $C(-\infty, \infty)$, the space or real valued continuous functions on $(-\infty, \infty)$.

Appendix 4A contains a brief account of the standard techniques that are

becoming well known in applied probability. For a detailed account refer

Kushner and Clark [K10],Chapter 3], Billingsly [B5], Parthasarathy [P1],

Iglehart [I1].

Theorem 4.3: Let P(k) be defined by (4.1) and assume (C4.1), (C4.2) and

(C4.3'). Then

    (a) $P^k(.)$ is tight on $C(-\infty, \infty)$ and the limit P(t) of any weakly-

           convergent subsequence satisfies

$$\frac{d\ P(t)}{dt} = W[P(t)]$$

(4.42)

           on $(-\infty, \infty)$

    (b) $\lim_{k \to \infty} P(k) = \beta_2$ in probability where recall that $S = \{\beta_2\}$ is the

           only invariant set of the above differential equation in the

           simplex $S_M(M=2)$.

The proof of this theorem is given in appendix 4A after a brief discussion on weak convergence techniques. In the statement of the theorem 4.3 and elsewhere the phrase "$\{P^k(.)\}$ is tight on $c(-\infty, \infty)$" is often used in place of the following correct one: "The sequence of probability measures induced by $P^k(.)$ on $C(-\infty, \infty)$ is tight". The major burden of the proof is in showing the <u>tightness</u> of the sequence of measures. Tightness implies that every subsequence of probability measures has a further subsequence which weakly converges to a probability measure. From this we can easily characterize the properties of the process $P(t)$ corresponding to the limiting probability measure.

<u>Remark 4.2</u>: As a consequence of Theorem 4.3 we get

$$\eta^* \triangleq \lim_{k \to \infty} \eta(k) = \lim_{k \to \infty} [P(k) \, d_1 + (1-P(k)) \, d_2]$$

$$= \beta_2 \, d_1 + (1-\beta_2) \, d_2 \text{ in probability}$$

$$> \frac{d_1 + d_2}{2} \text{ in probability} \tag{4.49}$$

Since $\beta_2 \gtrless \frac{1}{2}$ when $d_1 \gtrless d_2$. (refer (4.4).

Similarly, by the same argument leading to (4.20), for every $\varepsilon > 0$

$$\lim_{k \to \infty} | \, \eta(k) - \max (d_1, d_2) \, | < \varepsilon \text{ in probability} \tag{4.50}$$

Comparing the definitions (1.7) and (4.21) we see that (4.49) is an intermediate result between expediency and strong-expediency. Similarly (4.50) is stronger than ε-optimality defined in (1.9) but weaker than strong ε-optimality defined in (4.22).

4.4 Comments and Historical Remarks:

The method of analysis presented in section 4.2 is now classic and is well known in the literature on stochastic approximation [B7] [W1] [G3] [C10]. For other related work on time varying learning algorithms of the type discussed here refer Tsypkin and Poznyak [T6] Poznyak [P4].

Kushner's work [K11] entitled "On the general convergence results for stochastic approximations via weak convergence theory" is perhaps the earliest known to the author to deal with stochastic approximation algorithms without requiring square summability of step length parameter sequence. The recent monograph by Kushner and Clark [K10] describes the compactness method in great detail - constrained as well as unconstrained non-linear stochastic optimization problem for convergence with probability one and weak convergence with the weakest condition on the noise process.

It is interesting to note that the differential equation (4.42) characterizing the asymptotic properties of $P^0(t)$ obtained by compactness methods is the same as the one obtained in Chapter 2 (refer equation 2.13) by a method due to Norman [N14] for approximating the evolution of the mean of the process $\{P(k)\}$.

The idea of characterizing the limiting behavior of a random sequence (generated by an iterative procedure of the type 4.1) by analyzing the qualitative behavior of certain associated ordinary differential equation has been independently developed by various authors. Norman [N14] for the analysis of slow Markov learning in the context of mathematical psychology, Kurtz [K6] in connection with a study of jump Markov processes which have wide application in biology, ecology, radio-active disintegration, etc., Meerkov [M5] in the context of certain class of cooperative games with incomplete information called Goore-games Ljung [L12] and Kushner [K12]

for the analysis of general stochastic approximation that are often used

for the problem of identification and control of linear system with unknown

parameters.

4.5 APPENDIX

WEAK CONVERGENCE OF PROBABILITY MEASURES - AN INTRODUCTION:

The theory of weak convergence of probability measures has found, in
recent years, extensive applications to queueing theory [I1] optimal stochastic control problems [K13] identification and control of linear systems
with unknown parameters [K12] solutions to elliptic and parabolic partial
differential equations [K13] constrained and unconstrained non-linear
stochastic optimization techniques [K10] and not to mention the countless
applications in statistics and probability theory [I1] [B5]. The reason for
such widespread interest and application is that this theory provides fundamental tools in the study of "approximations." The approximations, as a rule,
involve much simpler processes and "converge" in the limit to the desired
process. In our example, analysis of the behavior of the left shifts of the
interpolated process, $\{P^k(t)\}_{k \geq 0}$ provide necessary information regarding
the asymptotic behavior of the sequence $\{P(k)\}_{k \geq 0}$.

Weak convergence is a generalization of the concept of "convergence in
distribution" to abstract valued random variables-that is, to random variables taking values in a complete, separable, metric space. As the sample
paths of the left-shifted, interpolated process $\{P^k(t)\}$ are real valued
continuous functions on $(-\infty, \infty)$, a natural choice for our purposes is the
space of $C(-\infty, \infty)$ valued random variables. It is well known that $C(-\infty, \infty)$,
the space of real valued continuous functions on $(-\infty, \infty)$ is a complete,
separable metric space under uniform convergence in compact intervals [D3].
The main advantage of dealing with the entire process instead of the value
of the process at fixed time is that weak convergence results for the func-
tionals would be of greater interest than the original process itself.

Before stating the conditions for weak convergence of probability measures
on $C(-\infty, \infty)$ let us first review the classical case of weak convergence in
real-line.

1. Weak convergence in real-line:

Consider the real line R^1 with the ordinary Euclidean metric and probability measures on the class of Borel sets denoted by R^1. It is well known
[L13] that this probability measure is completely determined by its distribution function[*] $F(.)$ where

$$P(-\infty, X] = F(X) \tag{4A.1}$$

<u>Definition 4A.1</u>: A sequence of distribution functions $\{F_n(.)\}_{n \geq 0}$ converge
<u>weakly</u> to a distribution function $F(.)$ (denoted by $F_n \Rightarrow F$) if

$$\lim_{n \to \infty} F_n(X) = F(X) \tag{4A.2}$$

at all continuity points of X

From (4A.1) and (4A.2) we have the following equivalent definition [B5].
Let δA refer to the <u>boundary</u> of a sub-set A of the real line R^1. Let
$\{P_n\}_{n \geq 0}$ be the sequence of probability measures corresponding to the
sequence $\{F_n(.)\}_{n \geq 0}$.

<u>Definition 4A.2</u>: The sequence of probability measures $\{P_n\}$ on (R^1, R^1)
converge weakly to a probability measure $P(P_n \Rightarrow P)$ if

$$\lim_{n \to \infty} P_n(A) = P(A) \tag{4A.3}$$

for all Borel sets A such that $P(\delta A) = 0$.

Let $BC[R^1]$ denote the space of real valued bounded continuous functions
on R^1. The following theorem is an immediate consequence of the above

---

[*] A real valued right-continuous, non-decreasing function, $F(.)$ is called a
probability distribution function if $\lim_{X \to \infty} F(X) = 1$ and $\lim_{X \to \infty} F(X) = 0$.
In this appendix by distribution function we refer to the probability
distribution function.

definitions (theorem 2.1 in [B5]).

Theorem 4A.1: <u>A sequence of probability measures</u> $\{P_n\}_{n \geq 0}$ <u>on</u> (R^1, R^1) <u>con-</u>
<u>verges weakly to a probability measure if</u> and only if

$$\lim_{n \to \infty} \int_{R^1} f(X) \, d \, P_n(dX) = \int_{R^1} f(X) \, d \, P(X) \qquad (4A.4)$$

<u>for all</u> $f(.) \in BC[R^1]$

In the literature (4A.4) is very often given as the definition of weak
convergence since it is easily extendable to abstract spaces.

For proving the uniqueness of the limit we need (theorem 1.3 in [B5]):

Lemma 4A.1 <u>Two probability measures P and Z on</u> (R^1, R^1) <u>coincide if</u>

$$\int_{R^1} f(X) \, d \, P(dX) = \int_{R^1} f(X) \, Q(dX) \qquad (4A.5)$$

<u>for</u> $f(.) \in C[R^1]$, <u>the space of real valued continuous functions on</u> R^1.

An immediate consequence of the theorem 4A.1 and Lemma 4A.1 is the
following

Corollary 4A.1: <u>The sequence</u> $\{P_n\}$ <u>cannot converge weakly to two different</u>
<u>limits at the same time.</u>

The following theorem called "Helly Selection Theorem" is quite fundamental [L13].

Theorem 4A.2: <u>For every sequence</u> $\{F_n(.)\}$ <u>of distribution functions there</u>
<u>exists a subsequence</u> $\{F_{n_i}(.)\}$ <u>and a non-decreasing, right continuous function</u>
<u>F(.) such that</u> $\lim_{i \to \infty} F_{n_i}(X) = F(X)$ <u>at all continuity points of X.</u>

Notice the function F(.) of the above theorem must satisfy $0 \leq F(X) \leq 1$
for all $X \in R^1$. If $\mu(.)$ is the measure corresponding to F(.) defined as

$$\mu(a,b] = F(b) - F(a) \qquad (4A.6)$$

then it is clear that $\lim_{X \to \infty} \mu(-\infty, X] \leq 1$. Thus $\mu(.)$ need not be a probability measure. If we can now ensure that $\mu(.)$ is a probability measure. Then the

above theorem guarantees the existence of a subsequence converging weakly to

a probability measure. In the following we work towards a condition that

will ensure that $\mu(.)$ is indeed a probability measure.

Definition 4A.3: A sequence of probability measures is said to be "tight"

if for every $\varepsilon > 0$ there exists a closed, bounded(compact)interval $K_\varepsilon = [a,b]$

such that

$$P_n(K_\varepsilon) > 1 - \varepsilon \text{ for all } n \qquad\qquad (4A.7)$$

In terms of the corresponding distribution functions, for every $\varepsilon > 0$ there

exists X and y such that $F_n(X) < \varepsilon$ and $F_n(y) > 1 - \varepsilon$ for all n. To under-

stand the meaning of this condition consider a sequence $\{F_n^1(.)\}$ such that

$F_n^1(X) = F_0^1(X-n)$ where $F_0^1(.)$ is a distribution function on the real-line.

Clearly, this sequence of functions violates the above condition. Thus

tightness condition prevents the "escape" of the probability mass to infinity.

In the following we shall say that a sequence of random variables is

tight if the corresponding probability measures is tight.

Definition 4A.4: A sequence of probability measures is said to be relatively

compact if every subsequence of it has a further subsequence that converge

weakly to a probability measure. (Compare this definition with theorem 4A.2).

With these preliminary definitions we now state the main theorem of the

weak convergence on the real line:

Theorem 4A.3: A sequence of probability measures $\{P_n\}$ in (R^1, R^1) is rela-

tively compact" if and only if it is "tight".

Remarks 4A.1: This theorem verbatim carries over to probability measures in

complete, separable metric spaces. Such an extension is called Prohorov's

theorem. In fact, the definition of weak convergence as given in (4A.4) and the concept of tightness and relative compactness of probability measure as stated above holds good in such general spaces as well. In applications it is the "if" part of the above theorem that is more useful. Accordingly, major effort will be to show that a sequence of probability measures is tight and then invoke Prohorov's theorem. For a thorough treatment of this generalization refer to Billingsly [B5].

The following theorem known as "continuous mapping theorem" is very useful in applications: Let h be a continuous real valued function defined on R^1.

Theorem 4A.4[*]: If $X_k \Rightarrow X$ then $h(X_k) \Rightarrow h(X)$.

As theorem 4A.3, this theorem also is true in the context of weak convergence of random variables taking values in complete, separable metric spaces.

2. Weak Convergence in $C[\alpha,\beta]$.

Under uniform convergence on finite intervals, it is well known [D3] that $C[\alpha,\beta]$ is a complete, separable metric space. Hence Prohorov's theorem is applicable to sequence of probability measures in this space if we can prove tightness. Recall that tightness is concerned with probability measures over a compact subset and compact subsets in $c[\alpha,\beta]$ are defined by Arzela-Ascoli theorem. The following theorem gives the conditions for tightness for a sequence of probability measures in $c[\alpha,\beta]$ and is a probabilistic version of the classical Arzela-Ascoli theorem [B5].

Theorem 4A.4: Let $\{X_k(.)\}$ be a sequence of stochastic processes with paths in $c[\alpha,\beta]$ with probability one. The sequence $\{X_k(.)\}$ is tight if and only if for each $\eta > 0$ there exists a real number $N_\eta < \infty$ such that

$$P[\ |X_k(\alpha)\ | \geq N\ ] \leq \eta \quad \text{for all } k = 1,2,\ldots \qquad (4A.8)$$

[*] $X_k \Rightarrow X$ means X_k converges weakly to X.

and for each ε > 0 and η > 0, there exists a δ ε(0,1) and a k_0 < ∞ such that

$$P \left\{ \begin{array}{c} \sup \\ |t-s|<\delta \\ \alpha \leq t,s \leq \beta \end{array} |X_k(t) - X_k(s)| \geq \epsilon \right\} \geq \eta \text{ for all } k \geq k_0 \quad (4A.9)$$

Refer Billingsly [B5] for a proof of this theorem. Notice (4A.8) refers to the uniform boundedness and (4A.9) refers to the equicontinuity of sample paths of the process $\{X_k(.)\}$. The following two conditions are comparatively simpler to verify then (4A.9) and also imply (4A.9).

If for every ε > 0, η > 0, there is a δ ε (0,1) and an integer k_0 such that

$$P \left[\begin{array}{c} \sup \\ s \leq t \leq s+\delta \end{array} |X_k(t)-X_k(s)| \geq \epsilon \right] \leq \eta \delta \quad (4A.10)$$

for all k ≤ k_0 and for all α ≤ s, s + δ ≤ β, then (A4.9) hold. The following is a sufficient condition for (4A.10) to hold. There exists a real L, a > 0 and b > 0

$$E [|X_k(t) - X_k(s)|^a] \leq K|t - s|^b \text{ for all } k \quad (4A.11)$$

Once tightness is proved all that remains is to identify the limiting probability measure P(.) or the corresponding limiting process X(.). One of the standard techniques for this is to show that the sequence of finite dimensional distributions of $\{X_k(.)\}$ also converge weakly to the finite dimensional distributions of X(.). In general finite dimensional distributions are easy to handle. As the probability measure of a stochastic process is uniquely defined by the finite dimensional distributions, P is uniquely determined by the weak limit of the finite-dimensional distributions of $\{X_k(.)\}$.

Notice that the condition (4A.11) for any fixed k is the Kolmogorov's criterion for path continuity of separable random process. As we are dealing with random processes which are continuous with probability one, it is interesting to note that we could use the above criterion (4A.11) to prove tightness of probability measures as well.

Proof of theorem 4.3: Recall that

$$P^k(t) = P^k(0) + F^k(t) + M^k(t) \qquad \text{for } t \geq t_k$$

$$= P(0) \qquad\qquad\qquad \text{for } t \leq t_k$$

where

$$F^k(t) = \int_0^t W[P^k(s)] \, ds$$

Our first task is to show that $P^k(t)$ is tight. To this end observe that $M^k(0) = 0$ and from (4A.8) we have $M^k(t) \to 0$ in probability as $k \to \infty$. Also (4A.8) imply the conditions (4A.8) and (4A.9) for tightness and $M^k(t)$ converges weakly to a function which is zero.

As $P^k(t)$ is bounded for s and k and $W(.)$ is continuous, $F^k(t)$ satisfy the conditions for tightness. Hence $P^k(t)$ is tight. Hence there exists a subsequence which has a further subsequence that converges weakly to a process $P(t)$ such that

$$P(t) = P(0) + \int_0^t W[P(s)] \, ds$$

$$\text{or} \quad \frac{dP(t)}{dt} = W[P(t)] \qquad\qquad (4A.12)$$

This proves the claim (a) of theorem (4.3).

To prove (b) define a function Q: $c[\alpha,\beta] \to$ Real line such that if $P(t) \, \varepsilon \, c[\alpha,\beta]$

$$Q\,[P(t)] = \sup_{-T \leq t \leq T} |\, P(t) - \beta_2 \,| \quad \text{for } T < \infty$$

Since $Q(.)$ is continuous on $c[\alpha,\beta]$, by the continuous mapping theorem $Q[P^k(t)]$ converges weakly to $Q[P(t)]$ as $k \to \infty$. As β_2 is the only asymptotically stable solution of (4A.12) we have for every $\varepsilon > 0$

$$\lim_{k \to \infty} P \ [\ \sup_{-T \le t \le T} \ | \ P(t) - \beta_2 | \ge \varepsilon] = 0 \qquad (4A.13)$$

from which (b) of theorem (4.3) follows. $\qquad$ Q.E.D.

4.6 Exercises:

4.1) Let a_k be a sequence of reals such that $a_k > 0$

$a_k \to 0$ as $k \to \infty$ and $\sum\limits_{k=0}^{\infty} a_k = \infty$.

If x_k is non-negative real sequence such that $\sum\limits_{k=0}^{\infty} a_k x_k < \infty$,

characterize the properties of the sequence x_k.

Hint: If $a_k = \dfrac{1}{k+1}$ and $\sum\limits_{k=0}^{\infty} a_k x_k < \infty$ implies x_k is of the form

$\dfrac{1}{(k+1)^{\epsilon}}$ for some $\epsilon < 0$

4.2) If x_k is a sequence of real-valued random variables such that

$\lim\limits_{k \to \infty} E[|x_k|] = 0$, then $\lim\limits_{k \to \infty} x_k = 0$ in probability.

4.3) If $\lim\limits_{k \to \infty} x_k = x$ in probability, then there exists a subsequence $\{x_{k_n}\}_{n \geq 0}$

such that $\lim\limits_{n \to \infty} x_{k_n} = x$ with probability one.

Hint: Examine the method of lemma 4.1.

4.4) Prove that $e(t)$ defined in (4.30) tends to zero uniformly for all t in

finite intervals.

4.5) Show that $\epsilon^k(t)$ defined in (4.34) tends to zero as $\sup\limits_{i \geq 0} \{\theta_i\}$ tends

to zero.

4.6) Show that (4.41) follows from (4.40).

PART II

APPLICATIONS

CHAPTER 5

TWO PERSON ZERO-SUM SEQUENTIAL STOCHASTIC GAMES WITH IMPERFECT and INCOMPLETE

INFORMATION - GAME MATRIX WITH SADDLE POINT IN PURE STRATEGIES

5.1 Introduction:

In this and the next chapter we present an application of the learning

algorithms developed in the previous chapters to two person zero sum games:

Let A and B be the two players. Both are allowed to use mixed strategies.

At any instant each player picks a pure strategy as a sample realization

from his mixed strategy. As a result of their joint action they receive a

random outcome which is either a success or failure. Since the game is a

zero-sum game A's success is B's failure and vice-versa. The following

assumptions are fundamental to our analysis: Either player has no knowledge

of the set of pure strategies available to the other player or the pure

strategy actually chosen by the other player at any stage of the game or the

distribution of the random outcome as a function of the pure strategies

chosen by them. Just based on the pure strategy chosen by him and the

random outcome he receives both the players individually update their mixed
strategies using a learning algorithm. This cycle continues and thus the
game is played sequentially. In short we consider a zero-sum game between
two players in which the players are totally decentralized, there is no
communication or transfer of information between them either before or
during the course of the play of the game and in fact they may not even
know that they are involved in a game situation at all. In this set-up our
aim is to find conditions on the learning algorithms such that both the
players in the long run will receive an expected payoff as close to the well
established game theoretic solutions (Von Neumann value) as desired.

Von Neumann and Morgenstern [V 7,pp30] distinguish between games with
complete information and those with incomplete information. The latter
differ from the former in the fact that either player lacks full information
about the rules of the game. For example, they may lack full information
about other players' or his own payoff functions, the set of strategies
available to other player or to themselves, etc. Another distinction which
has received considerable attention in the literature [H1] is between
games with perfect information and those with imperfect information depending
on whether or not each player has all the information about other players'
and his own previous choice of strategies and perhaps previous chance moves
by nature. Accordingly the game problem introduced in the previous paragraph
is justifiably called the two person zero-sum sequential stochastic games
with imperfect and incomplete information.

For the sake of easy analysis we divide the two person zero-sum games
into two disjoint sub-classes (1) game matrix with saddle point in pure stra-

tegies and (2) game matrix <u>without</u> saddle point in pure strategies. We have already seen that the learning algorithms also fall into two disjoint groups (1) ergodic and (2) absolutely expedient algorithms. Thus theoretically we have four different combinations of the game problem. Further using the machinery of chapter 4 we can consider applying time-varying analogues of the ergodic algorithms to the game problem as well.

In this chapter we present an analysis of perhaps the simplest case of the game matrix with saddlepoint in pure strategies and when the players use the L_{R-I}^A (a special case of the absolutely expedient, see chapter 3) learning algorithm. It is shown that there exists proper choice of parameters of the L_{R-I}^A learning algorithm (When both the players use the same or different values of the step length parameter) such that the expected payoff to either player can be made as close to the Von Neumann value of the game as desired. A number of simulation results are given and a variety of special cases are also discussed. As to be expected, the method of analysis to be used in this chapter is very similar to those developed in chapter 3.

In the sequential stochastic game that we consider here each player, ignorant of the fact that they are involved in a game, decide to use the learning algorithms that they would normally adopt in a game against nature. The fact that the same algorithms result in their converging asymptotically to optimal pure strategies of the zero-sum game when they exist is the unique feature of our result.

5.2   The L_{R-I}^A algorithm and statement of results:

Consider a zero-sum game between two players A and B in which A and B
have M and N pure strategies respectively. Suppose that at the kth play A
used a mixed strategy $p(k) = (p_1(k), p_2(k) \ldots, p_M(k)^T$ where $p_s(k)$ is the
probability of his choosing the sth pure strategy. Similarly B uses $q(k) =$
$(q_1(k), q_2(k), \ldots q_N(k)^T$. The outcome of the game depends on <u>chance</u> as
well as the choice of pure strategies by each player. There are only two
possible values for the outcome: +1 called <u>unit gain</u> for player A and –1
called <u>unit loss</u> for player A. The gain or loss for B is the same as loss
or gain for A. The game is played repeatedly and after each play either
player observes only a <u>random</u> <u>outcome</u>. Neither player is aware of the
mixed strategy used by the other player nor its pure realization. In fact,
each player does not even know how the distribution of the outcome is
determined as a function of the strategy pair or indeed what strategies
are available to the other player.

Using a learning algorithm each player increases the probability of
choosing a particular pure strategy if that strategy was chosen on the
previous play and led to a gain for him. The probabilities of all the other
pure strategies are then adjusted so that the total probability remains
unity. The specific algorithm used is the linear reward-inaction learning
algorithm of the absolutely expedient type described in detail in chapter 3.
Recall according to this algorithm the mixed strategies of either player
remain unchanged if they receive a loss.

The L_{R-I}^{A} algorithm:

Let "a" be a constant with $0 < a < 1$ (the learning parameter). At the kth play, let i be pure strategy actually used by player A, that is, i^{th} strategy is the sample realization of the mixed strategy $p(k)$. Similarly, let j be the strategy used by player B. Then the mixed strategies $p(k+1)$ and $q(k+1)$ to be used by A and B at the $(k+1)^{st}$ play is given by:

$$p_i(k+1) = p_i(k) + a[1 - p_i(k)] \qquad \text{if A received a unit gain}$$
$$\qquad\qquad\qquad\qquad\qquad\qquad\qquad \text{on the } k^{th} \text{ play} \qquad\qquad (5.1)$$
$$p_s(k+1) = p_s(k) - a\, p_s(k) \ (s \neq i)$$

$p_s(k+1) = p_s(k) \ s = 1,2,\ldots,M$, if A received a unit loss on the kth play.

$$q_j(k+1) = q_j(k) + a\, [1-q_j(k)] \qquad \text{if B received a unit gain on}$$
$$\qquad\qquad\qquad\qquad\qquad\qquad\qquad \text{the } k^{th} \text{ play} \qquad\qquad (5.2)$$
$$q_s(k+1) = q_s(k) - a\, q_s(k)$$

$q_s(k+1) = q_s(k) \ s = 1,2,\ldots,N$, if B received a unit loss on the kth play.

Denote the unit gain for A "success" and let d_{ij} be the probability of success if i and j are pure strategies chosen by the players A and B. Let $C_{ij} = 1 - d_{ij}$ be the probability of (unit) loss (to player A) for the same choice of pure strategies. It is assumed that chance acts independently on each play and $0 < d_{ij} < 1$ for all i and j. Let D be the (M x N) matrix of d_{ij}. Since A tries to increase and B tries to decrease the probability of success D is essentially the game matrix[1].

We make the following fundamental assumption that D has a saddle point in pure strategies and for convenience we take d_{11} to be the saddle point. For later use we further assume that

---

[1] The actual game matrix $G = [g_{ij}] = [2\, d_{ij}-1]$.

$$d_{11} < d_{12} \leq d_{13} \leq \ldots \ldots \leq d_{IN}$$

$$d_{11} > d_{21} \geq d_{31} \geq \ldots \ldots \geq d_{M1}$$

(5.3)

Suppose now that both players use the L_{R-I}^{A} algorithm (5.1) and (5.2) with arbitrary but fixed initial mixed strategies $p(0)$ and $q(0)$ where $0 < p_s(0) < 1$ and $0 < q_r(0) < 1$ for all s and r. The mixed strategies $p(k)$ and $q(k)$ as defined by (5.1) and (5.2) are random vectors and (randomly) depend on the previous outcomes and the choice of pure strategies. It can be seen that $\{(p(k), q(k))\}$ is a stationary Markov process (since the learning parameter a and the elements of the matrix D do not vary with time). However, the random process $\{p(k)\}$ and $\{q(k)\}$ considered individually are non-stationary Markov processes. Conditioned on the particular configuration of choices of pure strategies and outcomes on plays prior to the kth play, the probability of success at the kth play is $p^T(k) \, D \, q(k)$. The unconditional probability $\alpha(k)$ of success at the kth play is therefore given by

$$\alpha(k) = E[p^T(k) \, D \, q \, (k)]$$

(5.4)

where $E[.]$ is the expectation taken over all possible choices of pure strategies and outcomes prior to the kth play.

Statement of main result:

The principal result of this chapter is given by the following theorem 5.1

Theorem 5.1: For every $\varepsilon > 0$ there is an a^* with $0 < a^* < 1$ such that if the learning parameter a is less than a^* then

1) $\lim\limits_{k \to \infty} \alpha(k)$ exists

(5.5)

and

2) $\lim\limits_{k \to \infty} | \alpha(k) - d_{11} | < \varepsilon$

(5.6)

Stated in other words, if both players use the L_{R-I}^A algorithm with a sufficiently small parameter, the overall success probability eventually approaches the value of the game arbitrarily closely.

Remarks 5.1: Before we go on to analyze the game and prove the above theorem certain observations regarding our motivation to use L_{R-I}^A algorithm are in order. Let one of the players (say A) use an L_{R-I}^A algorithm while B always chooses a fixed pure strategy j. In this case A may be assumed to play a game against nature where his i^{th} action produces a success with probability d_{ij}. Since A uses the L_{R-I}^A algorithm, using the results of chapter 3, it follows that his asymptotic probability of success can be made as close to $\max_i \{d_{ij}\}$ ($\geq d_{11}$) as desired. In other words, when B uses a fixed pure strategy, the probability of success will be greater than or equal to d_{11}. Suppose now if B uses a fixed mixed strategy (say) $q = (\lambda_1, \lambda_2, \ldots \lambda_N)$, then again from chapter 3, the asymptotic probability of success is $\max_i (\sum_{j=1}^{N} \lambda_j d_{ij})$ and this value is greater than or equal to d_{11}. Hence, a fixed strategy (pure or mixed) on the part of either player results in an advantage to the other player. Thus in order that one player does not have an edge over the other, it is natural to assume that both players use the same L_{R-I}^A algorithm. (See also exercise 5.3)

Further L_{R-I}^A algorithm constitutes a simple prototype of the absolutely expedient algorithms. Its distance diminishing property has been well exploited to establish powerful convergence theorems (refer Norman [N10]) and we can use all these standard results for the analysis of the game problem.

5.3 Analysis of the game:

In this section we provide a proof of theorem 5.1 in a series of steps.

If S_L refers to the L-dimensional unit simplex, then from the definition it follows that $p(k) \in S_M$ and $q(k) \in S_N$ for all k. Clearly $S = S_M \times S_N$ is the state space of the stationary Markov process $\{(p(k), q(k))\}_{k \geq 0}$. If s is any point belonging to S where $s = (p,q)$, then the algorithm (5.1)-(5.2) can be written as

$$s(k+1) = T[s(k) \ (i(k),j(k)), \ e(k)] \tag{5.7}$$

where T: $S \times (I \times J) \times E \rightarrow S$. $i(k) \in I = \{1,2,\ldots,M\}$, $j(k) \in \{1,2,\ldots,N\} = J$, $e(k) \in E = \{$unit gain, unit loss$\}$ and $s(k) \in S$

The following theorem proves that the algorithm (5.1) - (5.2) is distance-diminishing (refer chapter 1 for the definition.).

<u>Theorem 5.2</u>: If the learning parameter $a \in (0,1)$ and elements of the matrix D are such that $d_{ij} \in (0,1)$ for all i and j, then the L_{R-I}^A learning algorithm (5.1) - (5.2) is distance diminishing.

<u>Proof</u>: Let $s_i(k) = (p^{(i)}(k), q^{(i)}(k))$ i = 1,2,

$$d[p^{(1)}(k), p^{(2)}(k)] = \left[\sum_{m=1}^{M} \left[p_m^{(1)}(k) - p_m^{(2)}(k) \right]^2 \right]^{1/2}$$

$$d[q^{(1)}(k), q^{(2)}(k)] = \left[\sum_{n=1}^{N} \left[q_n^{(1)}(k) - q_n^{(2)}(k) \right]^2 \right]^{1/2}$$

and

$$d[s_1(k), s_2(k)] = [d^2[p^{(1)}(k), p^{(2)}(k)] + d^2[q^{(1)}(k), q^{(2)}(k)]]^{1/2}$$

At stage k let A and B pick the pure strategies i and j respectively. Then

$$d[s_1(k+1),s_2(k+1)] = \begin{cases} [(1-a)^2 \, d^2[p^{(1)}(k),p^{(2)}(k)] + d^2[q^{(1)}(k),q^{(2)}(k)]]^{\frac{1}{2}} \\ \qquad\qquad \text{with probability } d_{ij} \\ [d^2[p^{(1)}(k),p^{(2)}(k)] + (1-a)^2 d^2[q^{(1)}(k),q^{(2)}(k)]]^{\frac{1}{2}} \\ \qquad\qquad \text{with probability } c_{ij} \end{cases}$$

Since $0 < a < 1$, $0 < d_{ij} < 1$ and $c_{ij} = 1 - d_{ij}$ from definition 1.3, chapter 1.
Theorem 5.2 follows immediately.

Q.E.D.

Remark 5.2: Our theorem 5.2 is a special case of the proposition 1, section 2.1, chapter 2 of Norman [N10].

Let V_M and V_N be sets corresponding to the vertices of the simplices S_M and S_N respectively, where V_M contains M-unit vectors (of dimension M) and V_N contains N unit vectors (of dimension N). The elements of V_M and V_N correspond to the pure strategies of the two players A and B. From (5.1) and (5.2) it follows that $p(k+1) = p(k)$ with probability one if and only if $p(k)$ ε V_M. Likewise $q(k+1) = q(k)$ with probability one if and only if $q(k)$ ε V_N. In other words, $V_M \times V_N$ constitute the set of all absorbing states of the Markov process $\{(p(k),q(k))\}_{k \geq 0}$. (See exercise 5.1).

The following definition is very basic and plays a crucial role in our analysis.

Definition 5.1: A Markov process associated with a distance diminishing learning algorithm with compact metric space as its state space is called COMPACT MARKOV PROCESS

Since $S = S_M \times S_N$ is a compact metric space (with Euclidean metric) and the algorithm (5.1) – (5.2) is distance diminishing, $\{(p(k),q(k))\}_{k \geq 0}$ is a compact Markov process. Recall that an absorbing state is an ergodic kernel (see exercise 1.1). Accordingly $V_M \times V_N$ constitute the set of all ergodic kernels of the compart Markov process $\{(p(k),q(k))\}_{k \geq 0}$.

Our next theorem shows that a compact Markov process, whatever be its initial state, is attracted to its ergodic kernels.

Theorem 5.3: Let (p, q) be any initial state of the compact Markov process $\{(p(k), q(k))\}_{k \geq 0}$ and let

$$\Gamma_{i\ j}[p,q] \triangleq \text{Prob}\left[\lim_{k \to \infty} d[(p(k),q(k)), (e_i^M, e_j^N)] = 0 \mid p(0)=p \ q(0)=q\right] \quad (5.8)$$

where e_s^L is the sth unit vector of dimension L and

$$V_M \times V_N = \{(e_i^M, e_j^N) \mid i = 1,2,\ldots,M \text{ and } j = 1,2,\ldots,N\}.$$

Then

$$\sum_{i=1}^{M} \sum_{j=1}^{N} \Gamma_{ij} [(p ,q)] = 1 \quad (5.9)$$

In other words the Markov process $\{(p(k),q(k))\}$ defined by the algorithm (5.1) - (5.2) converges with probability one to the set of all absorbing states $V_M \times V_N$.

A brief account of the theory of compact Markov processes along with the proof of theorem 5.3 is given in appendix 5.8.

Remark 5.3: An immediate corollary of theorem 5.3 is that $\alpha(k)$ (defined in (5.4)) converges with probability one, that is $\lim_{k \to \infty} \alpha(k)$ exists. Hence the first part of the theorem 5.1 is proved.

Let $(p^*,q^*) \in V_M \times V_N$ represent the state to which $(p(k),q(k))$ converges. Then (5.8) may be rewritten as

$$\Gamma_{ij} [(p,q)] = \text{Prob}[p^* = e_i^M, q^* = e_j^N \mid p(0) = p, q(0) = q] \quad (5.10)$$

Thus $\Gamma_{11}[(p,q)]$ denotes the probability with which both the players, as $k \to \infty$, will play their first (pure) strategies starting at any initial state (p,q). Since d_{11} is assumed to be the saddle point of the game matrix D, $\Gamma_{11}[(p,q)]$ in other words, is the probability with which both the players will pick their optimal pure strategies as $k \to \infty$. In the remainder of this section we consider the problem of computing $\Gamma_{11}[(p,q)]$.

Let D[S] (and $S = S_M \times S_N$) be the space of all real valued continuously differentiable functions with bounded derivatives on S. Let g[(p,q)] be an element of D[S]. The learning algorithm (5.1) - (5.2) defines an operator u as

$$u \, g[(p,q)] = E[g[(p(k+1), q(k+1))] \mid p(k) = p, \, q(k) = q] \qquad (5.11)$$

It follows from the definition that the operator u is linear and preserves positive functions.

The following theorem 5.4 characterizes $\Gamma_{11}[(p,q)]$.

Theorem 5.4: $\Gamma_{11}[(p,q)]$ is the only solution in D(S) of the equation

$$u \, \Gamma_{11}[(p,q)] = \Gamma_{11}[(p,q)] \qquad (5.12)$$

with the boundary conditions

$$\left. \begin{array}{l} \Gamma_{11}[(e_1^M, e_1^N)] = 1 \\ \Gamma_{11}[(e_i^M, e_j^N)] = 0 \quad i \neq 1, \quad j \neq 1 \end{array} \right\} \qquad (5.13)$$

Theorem 5.4 is the same as proposition 3.1 (chapter 3) and hence the proof of theorem 5.4 follows from Appendix 3.9 with obvious changes in notation. In fact much of what follows in this section parallel the development in chapter 3, especially sections 3.4 and 3.5. (See exercise 5.2)

Solution to equation (5.12) is, in general, not tractable and we attempt to obtain in the following a lower bound (Why?) $f_1[(p,q)]$ for $\Gamma_{11}[(p,q)]$ and show that $f_1[(p,q)]$ may be made as close to 1 as desired.

Recall (from definition 3.1) that a real valued function f on S is defined as

$$\left. \begin{array}{llll} \text{Subregular} & \text{if} & f[(p,q)] \leq u \, f[(p,q)] \\ \quad \text{regular} & \text{if} & f[(p,q)] = u \, f[(p,q)] \end{array} \right\} \qquad (5.14)$$

and

$$\left. \begin{array}{llll} \text{Super-regular} & \text{if} & f[(p,q)] \geq u \, f[(p,q)] \end{array} \right\}$$

Thus $\Gamma_{11}[(p,q)]$ satisfying (5.12) is a regular function and the following theorem 5.5 gives the relation between super, sub and regular functions.

Theorem 5.5: If $f[(p,q)] \in D[S]$ and is such that

$$f[(e_1^M, e_1^N)] = 1, \quad f[(e_1^M, e_j^N)] = 0 \quad i \neq 1, \; j \neq 1, \tag{5.15}$$

then

$$f[(p,q)] \geq \Gamma_{11}[(p,q)] \text{ if it is super regular}$$

and

$$f[(p,q)] \leq \Gamma_{11}[(p,q)] \text{ if it is subregular.}$$

Proof: Follows from the proof of Proposition 3.2 (Chapter 3)   Q.E.D.

In view of this theorem 5.5, our task reduces to one of finding a sub-regular function satisfying (5.15). To this end, consider

$$f_1[p,q)] = \frac{1 - e^{-\frac{x}{a} p_1 q_1}}{1 - e^{-\frac{x}{a}}}, \quad x > 0 \text{ and real} \tag{5.16}$$

where x is to be chosen. $f_1[(p,q)] \in D(S)$ and satisfies (5.15). Notice the r.h.s. of (5.16) depends only on the first component of the probability vectors p and q. Since super and subregular functions are closed under addition and multiplication by a positive constant and if $f[(p,q)]$ is super regular then $-f[(p,q)]$ is subregular, it follows that $f_1[(p,q)]$ in (5.16) is subregular if

$$f_2[(p,q)] \triangleq e^{-\frac{x}{a} p_1 q_1} \tag{5.17}$$

is super regular. We now determine conditions under which $f_2[(p,q)]$ is super regular.

From the definition of the operator u given in (5.11) and using (5.1) − (5.2) it can be shown that

$$u \, f_2[(p,q)] - f_2[(p,q)] = x \, p_1 \, q_1 \, G[x,p,q] \, f_2 \, [(p,q)] \qquad (5.18)$$

where

$$G[x, p, q] \triangleq - V[-x(1-p_1) \, q_1] \, \{(1-p_1) \sum_{k=1}^{N} \, q_k \, d_{1k}\}$$

$$- V[-x(1-q_1) \, p_1] \, \{(1-q_1) \sum_{j=1}^{M} \, p_j \, c_{j1}$$

$$+ V[x \, p_1 \, q_1] \left\{ \sum_{j=1}^{M} \sum_{k \neq 1}^{N} \, p_j \, q_k \, c_{jk} + \sum_{j \neq 1}^{M} \sum_{k=1}^{N} \, p_j q_k d_{jk} \right\} \qquad (5.19)$$

where (as in chapter 3)

$$V[y] = \begin{cases} \dfrac{e^y - 1}{y} & \text{if } y \neq 0 \\ \\ 1 & \text{if } y = 0 \end{cases} \qquad (5.20)$$

From the definition it follows that if

$$G[x, p, q] \leq 0 \qquad (5.21a)$$

for all p_i, $q_j \, \varepsilon \, (0,1)$, $i = 1$ to M, $j = 1$ to N and $x > 0$, then $f_2[(f,q)]$ would be a super regular function. The inequality (5.21a) can be rewritten as follows:

$$\left\{ \sum_{j=1}^{M} \sum_{k \neq 1}^{N} \, p_j \, q_k \, c_{jk} + \sum_{j \neq 1}^{M} \sum_{k=1}^{N} p_j \, q_k \, d_{jk} \right\}$$

$$\leq \frac{V[-x(1-p_1)q_1]}{V[x \, p_1 \, q_1]} \left\{ (1-p_1) \sum_{k=1}^{N} a_k d_{ik} \right\}$$

$$+ \frac{V[-x(1-q_1)p_1]}{V[x \, p_1 \, q_1]} \left\{ (1-q_1) \sum_{j=1}^{M} p_j c_{j1} \right\} \qquad (5.21b)$$

Lemma 5.1: For any $x > 0$ <u>and for all</u> p_1, $q_1 \, \varepsilon \, (0,1)$

1) $\dfrac{1}{V[x]} \leq \dfrac{V[-x(1-p_1) \, q_1]}{V[x \, p_1 \, q_1]}$

and $\qquad\qquad\qquad\qquad\qquad\qquad\qquad\qquad\qquad (5.22)$

2) $\dfrac{1}{V[x]} \leq \dfrac{V[-x(1-q_1)p_1]}{V[x \, p_1 \, q_1]}$

Proof: Define $H[y] = \ell n\, V[y]$ where ℓn refers to the natural logarithm. It follows $\frac{d\,H(y)}{dy} > 0$ and $\frac{d\,H(y)}{dy}$ is strictly increasing function of y (refer (3.41), (3.42) and (3.43)).

Let

$$g[x, p, q] = \frac{V[-x(1-p_1)q_1]}{V[x\, p_1\, q]}$$

and

$$\Delta[x, p, q] = \ell n\, g[x, p, q]$$

It follows that

$$\Delta[x, p, q] = H[-x(1-p_1)q_1] - H[x\, p_1\, q_1]$$

$$= -\int_{-x(1-p_1)q_1}^{x\, p_1 q_1} [\frac{d\,H[y]}{dy}]\, dy \tag{5.23}$$

As $|\, x\, p_1\, q_1\, | + |\, -x(1-p_1)\, q_1\, | \leq x$ and $\frac{d\,H[y]}{dy}$ is strictly increasing, (5.23) implies

$$\Delta[x, p, q] \geq -\int_0^x [\frac{d\,H[y]}{dy}]\, dy \tag{5.24}$$

Taking exponentials on both sides of (5.24) and remembering $H(0) = 1$, we obtain from (5.24) that

$$\frac{1}{V[x]} \leq g[x, p, q]$$

The second part of the lemma 5.1 follows likewise and the proof is complete.

Q.E.D.

In view of the lemma 5.1, the inequality (5.21b) is true if

$$\frac{1}{V[x]} \geq \frac{\sum\limits_{j=1}^{M} \sum\limits_{k\neq 1}^{N} p_j\, q_k\, c_{jk} + \sum\limits_{j\neq 1}^{M} \sum\limits_{k=1}^{N} p_j\, q_k\, d_{jk}}{(1-p_1) \sum\limits_{k=1}^{N} q_k\, d_{1k} + (1-q_1) \sum\limits_{j=1}^{M} p_j\, c_{j1}} \tag{5.25}$$

$$= \frac{p^T R q}{p^T Q q} \tag{5.26}$$

where

$$R = \begin{bmatrix} 0 & c_{12} & c_{13} & \cdot & \cdot & \cdot & \cdot & c_{in} \\ d_{21} & 1 & 1 & \cdot & \cdot & \cdot & \cdot & 1 \\ d_{31} & 1 & 1 & \cdot & \cdot & \cdot & \cdot & 1 \\ \cdot & \cdot & \cdot & & & & & \cdot \\ \cdot & \cdot & \cdot & & & & & \cdot \\ \cdot & \cdot & \cdot & & & & & \cdot \\ \cdot & \cdot & \cdot & & & & & \cdot \\ d_{M1} & 1 & 1 & \cdot & \cdot & \cdot & \cdot & 1 \end{bmatrix}$$

and

$$Q = \begin{bmatrix} 0 & c_{11} & c_{11} & \cdot & \cdot & \cdot & \cdot & c_{11} \\ d_{11} & c_{21}+d_{12} & c_{21}+d_{13} & \cdot & \cdot & c_{21}+d_{1N} \\ d_{11} & c_{31}+d_{12} & c_{31}+d_{13} & \cdot & \cdot & c_{31}+d_{1N} \\ \cdot & & & & & \cdot \\ \cdot & & & & & \cdot \\ \cdot & & & & & \cdot \\ \cdot & & & & & \cdot \\ d_{11} & c_{M1}+d_{12} & c_{M1}+d_{13} & \cdot & \cdot & c_{M1}+d_{1N} \end{bmatrix}$$

From (5.3) it follows that every non-zero element of the matrix Q is greater than the corresponding element of the matrix R and there exists a constant δ such that

$$\delta = \max_{\substack{i,j \\ i,j \neq 1}} \left\{ \frac{R_{ij}}{Q_{ij}} \right\} < 1 \qquad (5.27)$$

Hence inequality (5.26) is satisfied if

$$\frac{1}{V[x]} \geq \delta \quad \text{where} \quad \delta < 1 \qquad (5.28)$$

Since $V[x]$ is a continuous function with $V[0] = 1$ and $V[x] \to \infty$ as $x \to \infty$, there exists an $x^* > 0$ such that

$$\frac{1}{V[x]} \geq \delta \quad \text{for all } x \in (0, x^*]$$

choosing a value $x = y = x^*$ we obtain that

$$f_1[(p,q)] = \frac{1 - e^{-\frac{y}{a} p_1 q_1}}{1 - e^{-\frac{y}{a}}} \qquad (5.29)$$

is a subregular function satisfying (5.15) and hence

$$f_1[(p,q)] \leq \Gamma_{11}[(p,q)] \leq 1 \qquad (5.30)$$

Proof of part (2) of theorem 5.1:

Given a constant $\delta_1 > 0$, it follows from (5.29) that there exists a constant $a^* < 1$ such that for all $0 < a \leq a^*$

$$1 - \delta_1 \leq f_1[(p,q)] \leq \Gamma_{11}[(p,q)] \leq 1 \qquad (5.31)$$

Since

$$\lim_{k \to \infty} \alpha(k) = \Gamma_{11}[(p,q)] d_{11} + \{1 - \Gamma_{11}[p,q)]\} h[p^*, q^*, d_{ij}] \qquad (5.32)$$

where $h[., ., .]$ is a uniformly bounded function of its arguments, from (5.31) and (5.32) we have

$$(1 - \delta_1) d_{11} \leq \lim_{k \to \infty} \alpha(k) \leq d_{11} + \delta_1 |h|$$

Hence

$$\lim_{k \to \infty} | \alpha(k) - d_{11}| \leq \delta_1 \max (|d_{11}|, |h|) < \epsilon$$

by proper choice of δ_1. This completes the proof of the main theorem 5.1.

Q.E.D.

It is interesting to note that the elements of the matrix R and Q depend only on the row and column of D containing the saddle point. Further (5.3) is crucial for our analysis. Before closing this section we give an alternate viewpoint of the algorithm (5.1) - (5.2). Define an M x N matrix $H(k) = [h_{ij}(k)]$ where $h_{ij}(k) = p_i(k) q_j(k)$, that is, $h_{ij}(k)$ is the probability that players A and B will pick their i^{th} and j^{th} pure strategies respectively. Thus $\sum\limits_{i=1}^{M} \sum\limits_{j=1}^{N} h_{ij}(k) = 1$ for all $k \geq 0$. At

stage k let the players pick the pure strategies m and n respectively. If

this play (m,n) results in a unit gain for player A, then (5.1) - (5.2)

can be rephrased as

$$h_{mj}(k+1) > h_{mj}(k) \qquad j = 1,2,\ldots,N$$

$$h_{\ell j}(k+1) < h_{\ell j}(k) \qquad \ell \neq m, j = 1,2,\ldots,N$$

If the same play results in a unit loss to player B, then

$$h_{jn}(k+1) > h_{jn}(k) \qquad j = 1,2,\ldots,N$$

$$h_{j\ell}(k+1) < h_{j\ell}(k), \qquad \ell \neq n, \; j = 1,2,\ldots,N.$$

In other words, the learning algorithm we consider updates a matrix of

probabilities in such a way that A's gain changes the rows and B's gain

changes the columns of H(k). The above analysis shows that by proper choice

of the parameters of the algorithm the probability with which $h_{11}(k)$ con-

verges to 1 can be made as close to unity as desired.

5.4 Special case: Dominance:

A special case of the game matrix with saddle point in pure strate-

gies deserves a special mention here. It is the case when the game matrix

exhibit dominance. The reason for such a special interest stems from the

fact that it is only in this case we can directly deal with the non-

stationary Markov processes $\{P(k)\}_{k \geq 0}$ and $\{q(k)\}_{k \geq 0}$ individually.

The following definition of dominance is quite standard [V 7]

Definition 5: A row (column) of the matrix D is said to be dominant if

every element of that row (column) is greater (less) than or equal to the

corresponding elements of all the other rows (columns).

Thus, if $d_{ij} \geq d_{kj}$ for all $j = 1,2,\ldots,N$ and $i \neq k$, it is row dominant (it is required that no two rows are identical). Similarly for column dominance. Row dominance implies that player A has a pure strategy which is preferable to all the other strategies. A similar preference exists for player B with column dominance. Further, if the game matrix is both row and column dominant it has a saddle point. Except for the 2 x 2 case, the existence of a saddle point does not, however, imply either row or column dominance. In the 2 x 2 case existence of a saddle point implies either row dominance or column dominance or both.

In the following analysis we shall continue to assume that d_{11} is the saddle point in D and $0 < d_{ij} < 1$ for all i and j. Let

$$d_{1j} > d_{mj} \quad m \neq 1, \ j = 1,2,\ldots,N \tag{5.33}$$

that is, we consider the case in which the first row of D is strictly dominant in the sense of (5.33). Let d_i^A and c_i^A be the average probability of unit gain and unit loss respectively when player A picks his action i. Then, for $i = 1,2,\ldots,N$,

$$d_i^A(k) = \sum_{j=1}^{N} d_{ij} \, q_j(k) \ \text{ and } \ c_i^A(k) = 1 - d_i^A(k) \tag{5.34}$$

Assumption (5.33) implies that there exists a $\delta > 0$ such that

$$d_1^A(k) > d_j^A(k) + \delta \quad \text{for all } j \neq 1 \text{ and } k \geq 0 \tag{5.35}$$

The following theorem summarizes the effect of dominance.

Theorem 5.6: Let $\{p(k)\}$ evolve according to (5.1) and let the matrix satisfy the condition (5.33). If $0 < p_1(0) < 1$, then for every $\varepsilon > 0$ there exists an $a^* < 1$ such that for all $a < a^*$

$$\text{Prob} \left[\lim_{k \to \infty} p_1(k) = 1 \right] > 1 - \varepsilon \tag{5.36}$$

<u>Proof</u>: The proof of this theorem will be presented in various steps.

<u>Step 1</u>: From (5.1) it can be seen that

$$E[p_1(k+1) - p_1(k) \mid p(k)] = a \, p_1(k) \sum_{j \neq 1} p_j(k) [d_1^A(k) - d_j^A(k)] \quad (5.37)$$

The following conclusions are immediate from (5.37).

(c_1) The right hand side of 5.37 is non-negative and vanishes only at $p_1(k) = 0$

or 1

(c_2) Since $\{p(k)\}$ is a (non-stationary) Markov process $\{p_1(k)\}$ is a non-

negative bounded submartingale. Hence $\lim_{k \to \infty} p_1(k) \triangleq p_1^*$ exists with

probability one. Further as $p_1(k) \leq 1$, we have $\lim_{k \to \infty} E[p_1(k)] = E[p_1^*]$

(c_3) From (5.1) it follows that 0 and 1 are the only absorbing barriers

for the process $\{p_1(k)\}$ and for all $0 < p_1(k) < 1$, $p_1(k+1) \neq p_1(k)$ with

probability one. Thus $\lim_{k \to \infty} p_1(k) = p_1^* = 0$ or 1 with probability one.

(c_4) Define $\quad p_2'(k) = \sum_{j=2}^{M} p_j(k) \quad$ (5.38)

Clearly $\{p_2'(k)\}$ is a non-negative super-martingale and $\lim_{k \to \infty} p_2'(k) = 1 - p_1^*$

= 0 or 1 with probability one.

<u>Step 2</u>: Consider the function $e^{-\frac{x}{a} p_1(k)}$. Then there exists an $x > 0$ such

that

$$E[\, e^{-\frac{x}{a} p_1(k+1)} \mid p(k) \,] \leq e^{-\frac{x}{a} p_1(k)} \quad (5.39)$$

The proof of (5.39) follows very closely the development in section 3.5

and we merely indicate the major steps.

Let $p(k) = p = (p_1, p_2, \ldots, p_N)^T$. Then using (5.1) it can be seen

that

$$E[e^{-\frac{x}{a} p_1(k+1)} \mid p(k) = p] - e^{-\frac{x}{a} p_1} = -x \, F[x,k,p] e^{-\frac{x}{a} p_1} \quad (5.40)$$

where

$$F[x,k,p] = p_1(1-p_1) \, d_1^A(k) \, V \, [-x(1-p_1)] - p_1 \sum_{j \neq 1} p_j \, d_j^A(k) \, V[x \, p_i]$$

$$V[z] = \frac{e^z - 1}{z} \quad \text{if } z \neq 0$$

$$= 1 \quad \text{if } z = 0$$

$F(x,k,p) \geq 0$ if the following inequality is true:

$$\frac{V[-x(1-p_1)]}{V[x \, p_1]} \geq \frac{\sum\limits_{j \neq 1}^{M} p_j \, d_j^A(k)}{(1-p_1) \, d_1^A(k)} \tag{5.41}$$

But (from section 3.5)

$$\frac{V[-x(1-p_1)]}{V[x \, p_1]} \geq \frac{1}{V[x]} \tag{5.42}$$

and from (5.35)

$$1 > \eta \triangleq \max_{j \neq 1} \; \sup_k \left\{ \frac{d_j^A(k)}{d_2^A(k)} \right\} \geq \frac{\sum\limits_{j \neq 1}^{M} p_j \, d_j^A(k)}{(1-p_1) \, d_1^A(k)} \tag{5.43}$$

Combining (5.42) and (5.43) we see that (5.41) is true if

$$V[x] = \frac{1}{\eta} \tag{5.44}$$

Let y be the solution of (5.44). Hence $F[y, k, p] \geq 0$ and (5.39) is proved.

Step 3: Consider the function

$$h[x,a,z] = \frac{e^{\frac{x}{a} z} - 1}{e^{\frac{x}{a}} - 1} \tag{5.45}$$

There exists an $x > 0$ such that

$$E[h[x,a,p_2'(k+1)] \, | p(k)] \leq h[x,a,p_2'(k)] \tag{5.46}$$

where $p_2'(k)$ is defined in (5.38).

Notice

$$h[x,a,p_2'(k+1)] = \frac{e^{\frac{x}{a}[1-p_1(k+1)]} -1}{e^{\frac{x}{a}} -1}$$

$$= \frac{e^{-\frac{x}{a}p_1(k+1)} -e^{-\frac{x}{a}}}{1-e^{-\frac{x}{a}}} \tag{5.47}$$

Taking conditional expectations on both sides of (5.47), from the basic properties of conditional expectations and (5.39), we immediately get (5.46) for x = y defined by (5.44).

Step 4: Since $h[x,a,z]$ in a continuous function of z from the conclusion (c_4) given above we get

$$\lim_{k \to \infty} h[x,a,p_2'(k)] = h[x,a,(1-p_1^*)] \tag{5.48}$$

with probability one.

Further $\left.\begin{array}{l} 0 < h[x,a,z] < 1 \quad \text{for all } 0 < z < 1 \\[2mm] h[x,a,0] = 0 \quad \text{and} \quad h[x,a,1] = 1 \end{array}\right\}$ (5.49)

Combining (5.48) and (5.49) we obtain

$$h[x,a,(1-p_1^*)] = (1 - p_1^*) \quad \text{with probability one} \tag{5.50}$$

Step 5: Taking expectations on both sides of (5.46) with x = y we obtain,

$$E[h[y,a\ p_2'(k)]] \geq E[h[y,a,p_2'(k+1)]] \tag{5.51}$$

for all $k \geq 0$. Consequently (since $h[x,a,z]$ is a bounded function of z)

$$\begin{aligned} h[y,a,p_2'(0)] &\geq \lim_{k \to \infty} E[h[y,a,p_2'(k)] \\ &= E[\lim_{k \to \infty} h[y,a,p_2'(k)] \\ &= [(1-p_1^*] \quad \text{from (5.50)} \end{aligned} \tag{5.52}$$

But for any fixed x > 0 and 0 < z < 1 it follows that

$$\lim_{a \to 0} h[x,\ a,\ z] = 0 \tag{5.53}$$

Since $0 < p_2'(0) < 1$, for any $\varepsilon > 0$, (5.53) implies that there exists an $a^* < 1$ such that for all $a < a^*$

$$\varepsilon > h[y,a,p_2'(0)] \geq E[(1 - p_1^*)] \qquad (5.54)$$

and the theorem is proved.

<u>Step 6</u>: As $p_1^* = 0$ or 1 with probability one we obtain from (5.54)

$$\text{Prob } [p_1^* = 1] > 1 - \varepsilon \qquad (5.55)$$

Q.E.D.

In other words, if there is a row dominance and if the player A uses algorithm (5.1), irrespective of the behavior of B, the player A will in the long run choose the pure strategy corresponding to the dominating row with probability as close to unity as desired. A similar result will apply to player B under column dominance. (See exercise 5.5).

Further, in a 2 x 2 game existence of the saddle point in pure strategies imply either row and/or column dominance and the theorem 5.6 immediately apply to this case as well.

5.5 Simulations:

In this section we present a number of simulation experiments illustrating the theory developed in the previous sections.

Example 5.1: Consider a game matrix D given below

$$D = \begin{bmatrix} .65 & .25 & .575 \\ .64 & .625 & \boxed{60} \\ .675 & .775 & .375 \end{bmatrix}$$

The circled element refer to the saddle point. With the value of a = 0.02 (learning parameter) in (5.1) - (5.2), twenty-five sample runs, with 3500 iterations per run, were made. The average value of $p^T(k)$ D q(k) (averaged over twenty-five sample runs) is given for various values of k in table 5.1. The value of $p_i(0) = q_i(0) = \frac{1}{3}$ for i = 1, 2, 3.

Table 5.1

k	$E[p^k(k)\ D\ q(k)]$
0	0.5739
100	0.5749
200	0.5752
300	0.5785
400	0.5673
500	0.5763
1000	0.5732
1500	0.5720
2000	0.5969
2500	0.5978
3000	0.5998
3500	0.5999

Example 5.2: Consider a 2 x 4 game matrix

$$D = \begin{bmatrix} \textcircled{-.2} & -.1 & 0 & .4 \\ -.4 & -.2 & .1 & .5 \end{bmatrix}$$

The saddle point is the circled element. With $p_1(0) = p_2(0) = \frac{1}{2}$ and $q_i(0) = \frac{1}{4}$ i = 1, 2, 3, 4 and for a = 0.02, twenty five samples were made and the (numerical) average value of $p^T Dq$ for various values of k are given Table 5.2.

Table 5.2

k	$E[\, p^T(k) \ d \ q(k)]$
0	0.0125
100	−0.0593
200	−0.1196
300	−0.1592
400	−0.1735
500	−0.1882
1000	−0.2075
1500	−0.2043
2000	−0.1978
2500	−0.2000
3000	−0.1978
3500	−0.1989

5.6   Comments and historical remarks:

Historically Von Neumann and Morgenstern [V7] in their classic work
developed the theory of games with complete information. Scarf and
Shapley [S 6] studied sequential games with imperfect information. Later
Harsanyi [H 1] proposed the Bayesian approach to the games of incomplete
information where the players are uncertain about some important parameters of the game situation such as the payoff functions, the strategies
available to other players, the information other players have about the game,
etc. But each player has a subjective probability distribution over the
possible alternate values the unknown parameters can take. Harsanyi has
shown that under certain conditions of consistency of the above probability distributions games of incomplete information are game-theoreti-
cally equivalent to certain games with complete information, called
"Bayesian Games".

Aumann and Maschler [A6] began the study of sequential games with
incomplete information. Their work was later followed by a number of
authors, Zamir [Z1], Ponssard and Zamir [P3], Kohlberg [K2]. In
almost all these works, it is assumed that the true payoff matrix is one
out of m(>1) possible matrices. To begin with a <u>chance move</u> chooses
(according to a known distribution) a payoff matrix which remain
unchanged throughout the duration of the game. After each stage, each
player receives some information depending only on the actual payoff
matrix chosen by chance and the pure strategies chosen by both players
at that stage. The average payoff each player receives is a function of
the amount of information that each player has about the unknown. The
limiting behavior of the average payoff per stage has been extensively

investigated in [A6] [Z1] [P3] [K2].

Another related body of literature on game theory is concerned with the method of "fictitious" play for sequential games. At each stage both the players are made to choose the pure strategies that yield the best results against all the past choices of their opponents. In 1950, Brown [B11] conjectured that the above procedure would converge to the value of the game when it is a zero sum game. This conjecture was later proved by Robinson [R6]. Since then a variety of algorithms have been discussed in this context by Sanghvi and Sobel [S2].

With the exception of the work by Suppes and Atkinson [S17] much of the papers on the learning approach to games with incomplete and imperfect information has been confined to the literature on systems theory. Krylov and Tsetlin [K5] initiated research on this class of games in U.S.S.R. A recent monograph by Tsetlin [T2] summarizes the work of Russian authors. Papers by Chandrasekaran and Shen [C2] Viswanathan and Narendra [V 4] and Crawford [C9] represent other related work on this topic. In spite of such widespread interest, the problem of finding conditions under which a given class of learning algorithm will converge to the well established game theoretic solutions has remained open until recently. Much of the development in this chapter follows [L6].

In the next chapter we shall take up the analysis of two person zero-sum games without saddle point in pure strategies using ergodic learning algorithms described in chapter 2. Now in closing this section we shall mention two major open problems connected with the topic of this chapter.

1) Prove theorem 5.1 when both the players use absolutely expedient algorithm instead of the L^A_{R-I} algorithm (5.1) - (5.2).

2) Compute the probabilities $\Gamma_{ij}(p)$ when both the players use the algorithm (5.1) - (5.2) but the underlying game matrix does not have a saddle point in pure strategies.

5.7 Appendix

Compact Markov Processes:

In this appendix we summarize some of the important properties of compact Markov processes that are relevant to our developments in chapter 5, especially theorem 5.3. For further details refer section 3.3 and 3.4 of chapter 3, Norman [N10].

Let (S,d) be a complete metric space with d as metric and E be a finite set. If f is a real valued function on S, define

$$m(f) = \sup_{s \neq s'} \frac{|f(s)-f(s')|}{d(s,s')}$$

where $|.|$ refers to the absolute value. If $m(f) < \infty$, then f is said to satisfy Lipschitz condition. Let K[.,.] be the transition function for a Markov process {s(k)} $k \geq 0$ in a compact metric space (S,d), that is,

$$K[s,A] = \text{Prob } [s(k+1) \in A \mid s(k) = s]$$

Let U be the linear operator on bounded measurable real valued functions on S defined by

$$U f(s) = \int_S K[s, dt] f(t)$$

$$= E [f(s(k+1)) \mid s(k) = s]$$

It is assumed that

(I) K[s,.] is a probability measure for each $s \in S$.

(II) K[.,A] is a Borel measurable function for each Borel set A

Let C[S] be the space of continuous real-valued function on S under norm

$$|f| = \max_{s \in S} |f(s)|$$

and let C [S] be the space of functions that satisfy Lipschitz condition. It is well known [N1] that CL[S] is a Banach space with respect to the norm

$$||f|| = m(f) + |f|.$$

Norman has shown (refer chapter 3, [N10] that if the Markov process {s(k)} is associated with a distance-diminishing learning algorithm, then the following conditions are true.

(III) U maps CL[S] to CL[S] and bounded with respect to the norm $||.||$, that is,

$$||U|| = \sup_{\substack{f \in CL \\ f \neq 0}} \frac{||U f||}{||f||} < \infty$$

and

(IV) There is an integer k and there are two real numbers r < 1 and r < ∞ such that

$$m(U^k f) \leq r\, m(f) + R|f|$$

for all $f \in CL[S]$ where U^k is the kth iterate of the operator U.

<u>Definition</u>: Any Markov process in a compact metric space that satisfies the condition (I) to (IV) is called a "<u>compact Markov process</u>"

A non-empty Borel set B ⊂ S is <u>stochastically closed</u> if K[s,B] = 1 for all s ∈ B. A borel set is called <u>ergodic kernel</u> if it is stochastically and topologically closed and if it has no proper subset with these properties. It can be shown that distinct ergodic kernels are disjoint (chapter 3 in [N10]).

One of the fundamental properties of the compact Markov process is that whatever be the initial state of the process, it is always attracted to its ergodic kernels and this is the contents of the following theorem.

<u>Theorem</u>: <u>If a compact Markov process has L ergodic kernels denoted by</u> E_1, E_2, ..., E_L, <u>where</u>

$$\Gamma_i(s) = \text{Prob}\,[\lim_{k \to \infty} d(s(k), E_i) = 0 \mid s(0) = s]$$

<u>then</u>

$$\sum_{i=1}^{L} \Gamma_i(s) = 1 \quad \underline{\text{for all }} s \in S.$$

<u>Remark</u>: This is the first part of the theorem 4.3 in Norman [N10] page 54-55.

5.8 Exercises

5.1) Show that $V_M \times V_N$ is the only set of absorbing barriers for the algorithm (5.1) – (5.2) if $0 < a < 1$ and $0 < d_{ij} < 1$ for all i and j.

5.2) Work out the details of the proof for theorem 5.4.

5.3) In the algorithm (5.1) – (5.2) it is assumed that players A and B use the same learning parameter. Extend the analysis given in section (5.3) to the case when players A and B use different learning parameters say a_A and a_B.

5.4) Plot

$$h[x, a, z] = \frac{e^{\frac{x}{a}z} - 1}{e^{\frac{x}{a}} - 1}$$

as a function of $z \in [0,1]$ for various values of $a \in (0,1)$ and for various values of $x > 0$.

5.5) If there is row and/or column dominance and if the players use the algorithm (5.1) – (5.2), then using the results of theorem (5.6) show that either player will, in the long run, receive an average payoff as close to the von Neumann value of the game as desired.

5.6) Using the results of Chapter 3 prove the claims made in the remark 5.1.

5.7) For the following matrices compute the lower bound on $\Gamma_{\ell,m}[p]$ where the pair (ℓ,m) refers to the position of the saddle-point.

(a)
$$D = \begin{bmatrix} .6 & .8 \\ .4 & .1 \end{bmatrix}$$

(b)
$$D = \begin{bmatrix} .65 & .25 & .575 \\ .64 & .625 & .60 \\ .675 & .775 & .375 \end{bmatrix}$$

5.8) Consider a 2 x 2 game matrix
$$D = \begin{bmatrix} a & b \\ c & d \end{bmatrix}$$

5.8) cont'd.

where $0 < a, b, c, d < 1$ and $a > \max(b,c)$ and $d > \max(b,c)$. This

matrix has no saddle point in pure strategies.

a) If P_{opt} and q_{opt} represent the optimal strategies for A and B

respectively, then show that

$$P_{opt} = \left(\frac{d-c}{\Delta} , \frac{a-b}{\Delta} \right) \text{ and } q_{opt} = \left(\frac{d-b}{\Delta} , \frac{a-c}{\Delta} \right)$$

where $\Delta = (a+d) - (b+c)$

b) Let player B use the fixed mixed strategy q_{opt}. If player A was

algorithm (5.1), then show that

1) $\{p_1(k)\}$ is a martingale

2) $\lim_{k \to \infty} p_1(k) \; \varepsilon \; \{0,1\}$ with probability one.

c) Under the assumptions in (b) compute the probability of $p_1(k)$

converging to 0 or 1.

CHAPTER 6

TWO PERSON ZERO-SUM SEQUENTIAL STOCHASTIC

GAMES WITH IMPERFECT AND INCOMPLETE

INFORMATION--GENERAL CASE

6.1 INTRODUCTION:

In this chapter we present a unified approach to two person zero sum

games with incomplete and imperfect information wherein the game matrix may

not always have a sadlle point in pure strategies. This is a natural exten-

sion of the problem of Chapter 5. Under the assumption that both players A

and B use the L_{R-P}^{E} learning algorithm with the same reward and penalty

parameters but the penalty parameter being very small compared to the reward

parameter, it is shown that the expected mixed strategy of either player can

be made, asymptotically, as close to the optimal strategy dictated by the

game theory as desired, irrespective of whether or not the game matrix has

a saddle point in pure strategies.

To bring out the key ideas we only present an analysis of 2 x 2 game

using L_{R-P}^{E} learning algorithm. Obvious extensions to (i) When players

use the L_{R-P}^{E} algorithms with different sets of reward and penalty para-

meters (ii) When the players use the general N_{R-P}^{E} algorithms of Chapter 2

and (iii) M x N (M, N $\geq$ 2) game matrix are left as exercises. A number of

simulation results are given.

6.2. L^E_{R-P} Algorithm:

Consider a zero sum game between two players A and B in which both the players have two pure strategies each. Suppose at kth play, player A uses a mixed strategy $p(k) = (p_1(k), p_2(k))^T$ where $p_s(k)$ is the probability of choosing the sth pure strategy. Similarly, player B uses $q(k) = (q_1(k), q_2(k))^T$. The outcome of the game depends on chance and the choice of pure strategies by both players. There are only two possible values the random outcome can take: +1 called unit gain and -1 unit loss and the gain or loss for A is the loss or gain for B. The game is played repeatedly and after each play both the players observe only the random outcome. At any stage both the players are not aware of the mixed strategy used by other player or its pure realization. In fact, they do not even know the distribution of the random outcome as a function of the strategy pair or indeed what set of strategies are available to the other player.

Using a learning algorithm each player increases (decreases) the probability of choosing a particular pure strategy if that strategy was chosen on previous play and resulted in an unit gain (loss). All the other probabilities are adjusted to keep the total probability equal to unity. The algorithm is given below: Let θ_i be a constant such that $0 < \theta_i < 1$. At the kth play, let the player A choose ith pure strategy and player B use the jth pure strategy as a sample realization from $p(k)$ and $q(k)$ respectively. Then $p(k + 1)$ defined by

$$P_i(k + 1) = P_i(k) + \theta_1[1 - P_i(k)]$$
$$P_s(k + 1) = P_s(k) - \theta_1 P_s(k), \quad s \neq i$$

if A received a unit gain on kth play

(6.1)

$$P_i(k + 1) = P_i(k) - \theta_2 P_i(k)$$
$$P_s(k + 1) = P_s(k) + \theta_2 P_i(k), \quad s \neq i$$

if A received a unit loss on kth play.

$q(k + 1)$ for player B is defined exactly in the same way but replacing p by q, i by j and A by B.

The pair (i,j) denote the play at any stage where i and j are the pure strategies chosen by A and B respectively. Also in the sequel by gain or loss we will refer to player A's gain or loss. For the play (i,j) let d_{ij} ($c_{ij} = 1 - d_{ij}$) be the probability of gain (loss) where it is __assumed__ that $0 < d_{ij} < 1$ for all i and j = 1,2. Let D denote the matrix of d_{ij}'s. As player A tries to increase the gain and player B tries to decrease the gain, D is essentially the game matrix[1].

When both the players use the algorithm (6.1), it can be easily seen that the process $\{(p(k), q(k))\}$ $k \geq 0$ is a stationary Markov process while $\{p(k)\}$ and $\{q(k)\}$ for $k \geq 0$ are individually non-stationary Markov processes. Let $0 \leq p_s(0) \leq 1$ and $0 \leq q_s(0) \leq 1$ for s = 1,2. Define

$$\eta(k) = E[p^T(k)]DE[q(k)] \qquad (6.2)$$

where $E[\cdot]$ is the expectation taken over all possible choice of pure strategies by either player and all the random outcomes prior to kth play. Let $\theta_1 = \theta a$ and $\theta_2 = \theta b$ where θ, a, b are positive real constants and V be the Von-Neumann value of the game (corresponding to the game matrix D). Our main result is summarized by the following:

___
[1] The actual game matrix $G = [g_{ij}] = [2d_{ij} - 1]$

<u>Theorem 6.1</u>: – <u>For any given</u> $0 < a < 1$ <u>and for every</u> $\varepsilon > 0$ <u>there exists</u> $0 < b^* < a$ <u>and</u> $0 < \theta^* < 1$ <u>such that for all</u> $0 < b < b^*$ <u>and</u> $0 < \theta < \theta^*$

$$\lim_{k \to \infty} \left| \eta(k) - V \right| < \varepsilon \qquad\qquad (6.3)$$

<u>Remark 6.1</u>: – It is easily seen by comparison with remark 2.4 that (6.1) is the L_{R-P}^E algorithm. The parameters a and b are called reward and penalty parameters respectively and θ essentially control the step size. Stated in words the above theorem asserts that if both the players use the L_{R-P}^E algorithm with a very small penalty term compared to reward term and with a small step size parameter, they can ε-achieve the Von-Neumann value of the game in the long run.

6.3. Analysis of Game:

Let $S \triangleq [0,1] \times [0,1]$, $E = \{\text{unit gain, unit loss}\}$ and $I = \{1,2\}$. Define $P(k) = (p_1(k), q_1(k))$. S is the state space of the Markov process $P(k)$ and E called the event space and the algorithm (6.1) itself can be rewritten as

$$P(k + 1) = T[P(k), (i(k), j(k)), e(k)] \qquad\qquad (6.4)$$

where T: S x (I x I) x E→S,   i(k) and j(k) are the pure strategies chosen by players A and B at time k and e(k) is the random outcome,
e(k) ε E   The following theorem is immediate:

Theorem 6.2: If $0 < d_{ij} < 1$ and $0 < \theta_i < 1$ for all i, j = 1,2, then all
the states of the Markov Process {p(k)} are non-absorbing.

Proof: Left as an exercise 6.3.   Q.E.D.

Thus {P(k)} is an indexed (by θ) family of non-absorbing Markov process that evolve under small steps when θ is small. Chapter 2 deals with
the study of this class of processes and much of this chapter follows the
analysis given in that chapter.

Define

$$\delta P(k) = (\delta p_1(k), \delta q_1(k))^T \tag{6.5}$$

where

$$\delta P_1(k) = P_1(k+1) - p_1(k) \text{ and similarly for } \delta q_1(k).$$

It follows from the algorithm (6.1) that

$$(P6.1) \quad E[\delta P(k)|P(k) = P \triangleq (p_1, q_1)^T] = \theta W(P)$$

$$W(P) = (W_1(P), W_2(P))^T$$

$$W_i(P) = W_i^R(P) + W_i^P(P) \quad i = 1,2$$

$$W_1^R(P) = a\, p_1(1 - p_1)[c_2^A(q_1) - c_1^A(q_1)]$$

$$W_2^R(P) = a\, q_1(1 - q_1)[c_2^B(p_1) - c_1^B(p_1)]$$

$$W_1^P(P) = b[(1 - p_1)^2 c_2^A(q_1) - p_1^2 c_1^A(q_1)]$$

$$W_2^P(p) = b[(1 - q_1)^2 c_2^B(p_1) - q_1^2 c_1^B(p_1)]$$

Where

$$\left. \begin{array}{l} c_i^A(q_1) = c_{i1}q_1 + c_{i2}(1 - q_1) \\[2mm] c_i^B(p_1) = d_{i1}p_1 + d_{i2}(1 - p_1) \end{array} \right\} \quad i = 1,2 \qquad (6.6)$$

Similarly,

(P6.2) $E[\{\delta P(k) - \theta W(P)\}^T \{\delta P(k) - \theta W(P)\} | P(k) = P] = \theta^2 s(P)$

where

$$s(\ ) = a(P) - W^T(P)W(P)$$

$$a(P) = E[\delta P^T(k)\delta P(k) | P(k) = P]$$

The elements of the matrix can be easily computed and as all the states of the Markov process are non-absorbing, it can be easily shown that s(P) is positive definite for all P ε S. (see exercise 6.4)

Further

(P6.3) $E[|\delta P^T(k)|^3 | P(k) = P] = 0(\theta^3)$ for all P ε S

where $|x|$ refers to the length of the vector x.

(P6.4) W(P) has a bounded Lipschitz derivative in S.

(P6.5) s(P) is Lipschitz in S.

Properties (P6.3) - (P6.5) can be easily verified using routine arguments. (see exercise 6.5).

Theorems 6.3 and 6.4 below summarize further properties of W(P).

<u>Theorem 6.3</u>: <u>There exists an unique</u> $P* = (p_1^*, q_1^*)^T$ ε S <u>such that</u> W(P*) = 0.

<u>Proof</u>: On rewriting $W_1(P)$ becomes

$$\left. \begin{array}{l} W_1(P) = - (a-b)\left\{ c_2^A(q_1) - c_1^A(q_1)\right\} p_1^2 \\[2mm] \qquad + \{a[c_2^A(q_1) - c_1^A(q_1)] - 2bc_2^A(q_1)\} p_1 \\[2mm] \qquad + bc_2^A(q_1) \end{array} \right\} \qquad (6.7)$$

For any fixed $q_1 \varepsilon [0,1]$; $W_1(P)$ is a quadratic in p_1 and

$$W_1(P) = bc_2^A(q_1) > 0 \quad \text{at } p_1 = 0$$

$$= -bc_1^A(q_1) < 0 \quad \text{at } p_1 = 1$$

Thus for a fixed q_1, there exists an unique state $P(q_1) \triangleq (p_1(q_1), q_1)^T$ (to emphasize the dependence on q_1) for which $W_1(P(q_1)) = 0$. Similarly, for any fixed $p_1 \varepsilon [0,1]$ there exists an unique state $P(p_1) \triangleq (p_1, q_1(p_1))^T$ for which $W(P(p_1)) = 0$. Rest of the proof follows that of theorems 2.1 and corollary 2.1 and the details are given in Appendix. It is shown there that the locus of $p_1(q_1)$ and $q_1(p_1)$ as p_1 and q_1 varies has an <u>unique</u> intersection point $P^* = (p_1^*, q_1^*)^T \varepsilon S$ such that $W(P^*) = 0$. Q.E.D.

Let $P_{opt} = (p_{opt}, q_{opt})^T$ be the state that corresponds to the optimal strategy for both players. For purposes of easy presentation in the following we distinguish two cases:

Case 1: The game matrix has no saddle point in pure strategies.

In this case without loss of generality let us assume

$$d_{11} > \max\{d_{12}, d_{21}\}$$

$$d_{22} > \max\{d_{12}, d_{21}\}$$

(6.8)

It can be shown [V7] (see also exercise 5.8) that

$$P_{opt} = \frac{d_{22} - d_{21}}{\Delta}$$

(6.9)

$$q_{opt} = \frac{d_{22} - d_{12}}{\Delta}$$

and

$$\Delta = (d_{11} + d_{22}) - (d_{12} + d_{21})$$

Case 2: The game matrix has a saddle point in pure strategies.

In this case we shall assume that

$$d_{21} < d_{11} < d_{12} \tag{6.10}$$

and as a consequence $p_{opt} = q_{opt} = 1.0$.

Remark 6.2: All the other relations between the elements of the matrix D can be reduced to (6.8) or (6.10) by suitable relabelling of players and/or strategies.

The following corollary is an immediate consequence of Theorem 6.3.

Corollary 6.1: For any value of the reward parameter "a" and a given $\delta > 0$, there exists $b^* < a$ such that

$$|P^* - P_{opt}| < \delta \text{ for all } 0 < b < b^* \tag{6.11}$$

where

$$W(P^*) = 0.$$

Proof: The proof of this corollary is closely related to the proof of Theorem 6.3 and is given in Appendix 6.7.   Q.E.D.

The following examples illustrate theorem 6.3 and corollary 6.1.

Example 6.1: Consider a D matrix

$$D = \begin{bmatrix} .60 & .20 \\ .35 & .90 \end{bmatrix}$$

The plot of $W_1(P) = 0$ and $W_2(P) = 0$ for various sets of values of a and b are given figures 6.1 through 6.3. Notice the above matrix D has no saddle point in pure strategies and $p_{opt} = 0.57895$ and $q_{opt} = 0.7368$.

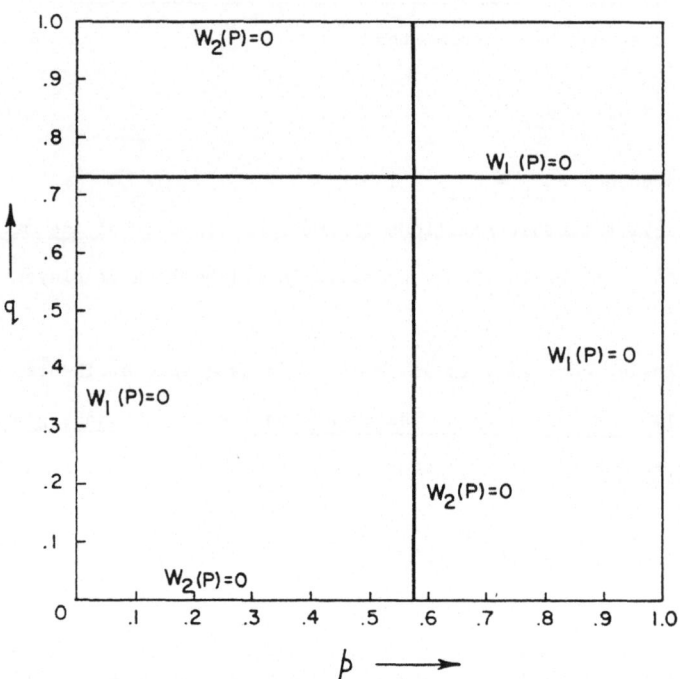

FIGURE 6.1

a = 0.20

b = 0.0

$$D = \begin{bmatrix} .60 & .20 \\ .35 & .90 \end{bmatrix}$$

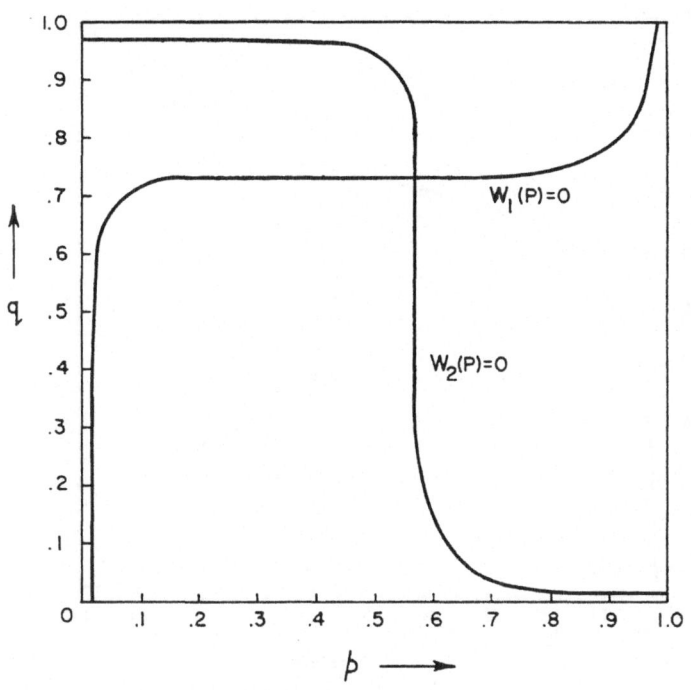

FIGURE 6.2

a = 0.200

b = 0.001

$$D = \begin{bmatrix} .60 & .20 \\ .35 & .90 \end{bmatrix}$$

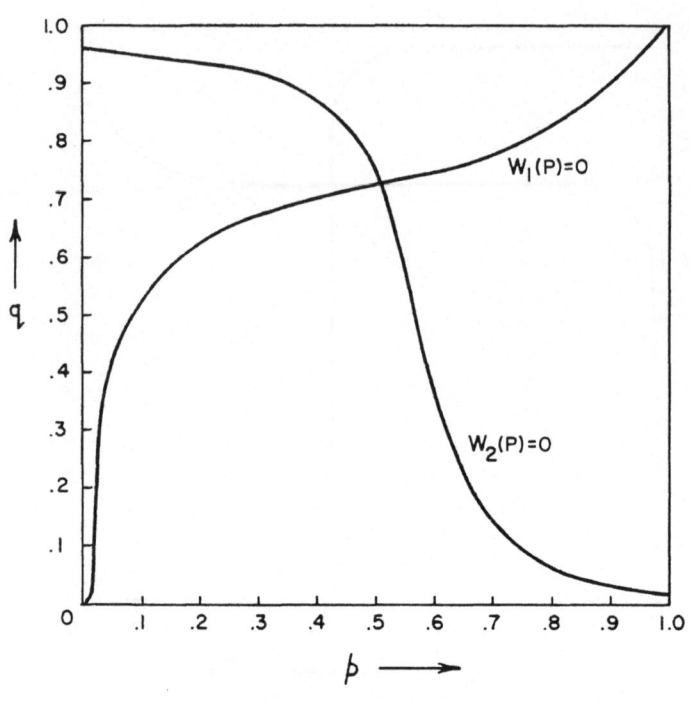

<u>FIGURE 6.3</u>

a = 0.20

b = 0.01

$$D = \begin{bmatrix} .60 & .20 \\ .35 & .90 \end{bmatrix}$$

Example 6.2:   For the matrix

$$D = \begin{bmatrix} .60 & .80 \\ .35 & .90 \end{bmatrix}$$

The plot of $W_1(P) = 0$ and $W_2(P) = 0$ are given in Figure 6.4. Also $p_{opt} = q_{opt} = 1.0$.

Denoting the Hessian of $W(P)$ as $W'(P)$, we obtain

$$W'(P) = \begin{vmatrix} \dfrac{\delta W_1(P)}{\delta p_1} & \dfrac{\delta W_1(P)}{\delta q_1} \\[3mm] \dfrac{\delta W_2(P)}{\delta p_1} & \dfrac{\delta W_2(P)}{\delta q_1} \end{vmatrix} \qquad (6.12)$$

where

$$\frac{\delta W_1(P)}{\delta p_1} = a[1-2p_1][c_2^A(q_1) - c_1^A(q_1)] - 2b[(1-p_1)c_2^A(q_1) + p_1 c_1^A(q_1)]$$

$$\frac{\delta W_1(P)}{\delta q_1} = a\Delta p_1(1-p_1) + b[(1-p_1)^2(d_{22} - d_{21}) - p_1^2(d_{12} - d_{11})]$$

$$\frac{\delta W_2(P)}{\delta p_1} = -a\Delta q_1(1-q_1) + b[(1-q_1)^2(d_{12} - d_{22}) - q_1^2(d_{11} - d_{21})]$$

and

$$\frac{\delta W_2(P)}{\delta q_1} = a[1-2q_1][c_2^B(p_1) - c_1^B(p_1)] - 2b[(1-q_1)c_2^B(p_1) + q_1 c_1^B(p_1)]$$

where Δ is given in (6.9).

Theorem 6.4 characterizes the properties of $W'(P)$.

Theorem 6.4: $W'(P*)$ is negative definite where $P*$ is as defined in Corollary 6.1.

Proof: Case 1:   From Corollary 6.1 it follows that by proper choice of the penalty parameter "b" (b $<<$ a) we can make $p_1^* \simeq p_{opt}$ and $q_1^* \simeq q_{opt}$. With such a choice of the parameter "b", it can be seen that

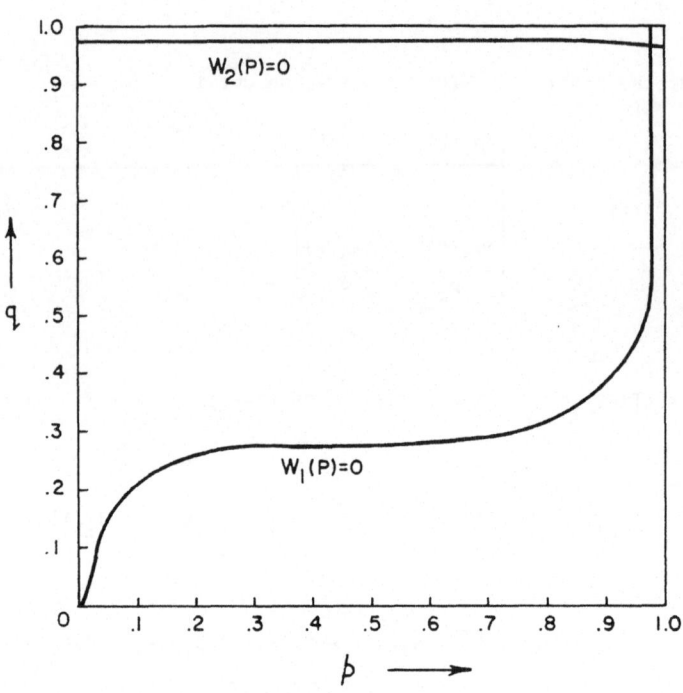

FIGURE 6.4

a = 0.20

b = 0.001

$$D = \begin{bmatrix} .60 & .80 \\ .35 & .90 \end{bmatrix}$$

$$c_i^A(q_{opt}) \simeq c_i^A(q_1^*) \text{ and } c_i^B(p_{opt}) \simeq c_i^B(p_1^*).$$

Further, in case 1

$$
\left.
\begin{aligned}
c_2^A(q_{opt}) &= c_1^A(q_{opt}) \\
\text{and} & \\
c_2^B(p_{opt}) &= c_2^B(p_{opt})
\end{aligned}
\right\}
\tag{6.13}
$$

Thus at the point $p^* = (p_1^*, q_1^*)$, $W'(P^*)$ can be well approximated by

$$
W'(P^*) \simeq
\begin{bmatrix}
-2bc_2^A(q_{opt}) & \dfrac{(a+b)}{\Delta}(d_{11} - d_{12})(d_{22} - d_{21}) \\
& \\
-\dfrac{(a+b)}{\Delta}(d_{11} - d_{21})(d_{22} - d_{12}) & -2bc_2^B(p_{opt})
\end{bmatrix}
$$

Let u_1 and v_1 denote the trace and determinant of $W'(P^*)$. Then

$$
\left.
\begin{aligned}
u_1 &= -2[c_2^A(q_{opt}) + c_2^B(p_{opt})] < 0 \\
v_1 &= 4^2 c_2^A(q_{opt}) c_2^B(p_{opt}) + \frac{(b+a)^2}{\Delta^2} F
\end{aligned}
\right\}
\tag{6.14}
$$

where

$$F = (d_{11} - d_{21})(d_{22} - d_{12})(d_{11} - d_{12})(d_{22} - d_{21}).$$

From (6.8) it follows that $F > 0$ and hence $v > 0$. Also

$$
u^2 - 4v = 4b^2[c_2^A(q_{opt}) - c_2^B(p_{opt})]^2 - 4\frac{(b+a)^2}{\Delta^2}F
\tag{6.15}
$$

$$
\simeq - \frac{4\,a(a+2B)F}{\Delta^2}
$$

$$< 0$$

From (6.15) it follows from [B6] that the eigenvalues of $W'(P^*)$ are complex conjugates with negative real parts.

Case 2: In this case it follows from corollary 6.1 that $p^* \approx 1$ and $q^* \approx 1$ and $W'(P^*)$ is such that

$$W'(P^*) \approx \begin{bmatrix} -a \; (d_{11} - d_{21}) - 2b(1-d_{11}) & -b \; (d_{12} - d_{11}) \\ \\ -b \; (d_{11} - d_{21}) & -a(d_{12} - d_{11}) - 2bd_{11} \end{bmatrix}$$

Using (6.10) we obtain

$$\left. \begin{aligned} u_1 &= -a \; [(d_{11}-d_{21}) + (d_{12}-d_{11})] - 2b < 0 \\ v_1 &\approx a^2(d_{11}-d_{21})(d_{12}-d_{11}) + 2ab[d_{11}(d_{11}-d_{21}) + (1-d_{11})(d_{12}-d_{11})] \\ &> 0 \end{aligned} \right\} (6.16)$$

But $u_1^2 - 4v_1$ is positive only for every small values of b. Again it follows from (6.16) and [23] that the eigenvalues of $W'(P^*)$ are real and negative or complex conjugates with negative real parts.

Thus in either case the theorem is true.   Q.E.D.

Notice the value of P^* and eigenvalues of $W'(P^*)$ are independent of the step length parameter θ.

The validity of the above approximations are illustrated by the following examples:

Example 6.1   (continued) : Table 6.1 gives the variation of the values of p_1^* and q_1^* and that of the eigenvalues λ_1 and λ_2 of the matrix $W'(P^*)$ as a function of the penalty parameter "b" for the game matrix D given in example 1.

Table 6.1

a	b	p^*	q^*	λ_1		λ_2	
				$\mathrm{Re}(\lambda_1)$	$\mathrm{Im}(\lambda_1)$	$\mathrm{Re}(\lambda_2)$	$\mathrm{Im}(\lambda_2)$
0.20	0.0	.5789	.7368	0.0	0.0413	0.0	-0.0413
0.20	0.001	.5655	.7384	-0.00131	0.0415	-0.00131	-0.0415
0.20	0.005	.5469	.7412	-0.00652	0.0424	-0.00652	-0.0424
0.20	0.01	.5169	.7400	-0.01280	0.0436	-0.01280	-0.0436
0.20	0.05	.4713	.7257	-0.02470	0.0463	-0.02470	-0.0463

Example 6.2: (continued) For the game matrix D given in example 2 table 6.2 gives the variation of the same quantities as in the above example as a function of "b".

Table 6.2

a	b	p^*	q^*	λ_1		λ_2	
				$\mathrm{Re}(\lambda_1)$	$\mathrm{Im}(\lambda_1)$	$\mathrm{Re}(\lambda_2)$	$\mathrm{Im}(\lambda_2)$
0.20	0.0	1.00	1.00	-0.0500	0.0	-0.0400	0.0
0.20	0.001	0.99196	0.98548	-0.0489	0.0	-0.0406	0.0
0.20	0.002	0.98385	0.97187	-0.0478	0.0	-0.0415	0.0
0.20	0.005	0.95932	0.93599	-0.0446	+0.00286	-0.0446	-0.00286
0.20	0.01	0.91885	0.88954	-0.0455	+0.00518	-0.0455	-0.00518

All the calculations leading to tables 6.1 and 6.2 are exact up to the double precision arithmetic in IBM 370/155. These examples help illustrating the accuracy of the approximations used in the proof of theorem 6.4.

The following is an analogue of theorem 2.3 and is an immediate consequence of the properties of W(P) given by theorems 6.3 and 6.4.

Theorem 6.5: Let both the players A and B use the L_{R-P}^E learning algorithm (6.1) with $\theta_1 = \theta a$ and $\theta_2 = \theta b$. If

a) there exists an unique $P^* \epsilon$ S such that $W(P^*) = 0$ and

b) $W'(P^*)$ is negative definite, then

$$E[P(k) \mid P(0) = P = (p,q)^T] = f(k\theta) + 0(\theta) \qquad (6.17)$$

uniformly for all $k \geq 0$ and $P \epsilon$ S where $f(t)$ is the solution of the following ordinary first order differential equation

$$f'(t) = W(f(t)) \quad \text{with } f(0) = P \qquad (6.18)$$

Proof: Same as theorem 2.3 and we invite the reader to fill in the details
(see exercise 6.6)

Q.E.D.

Remark 6.3: $E[P(k) \mid P(0) = P]$ is called the "mean learning" curve. Theorem 6.5 states that the mean learning curve can be well approximated by the solution of the differential equation (6.18) and the accuracy of the approximation is essentially controlled by the step length parameter θ.

We now turn to prove theorem 6.1.

Proof of Theorem 6.1: -

Properties of W (P) imply that the differential equation (6.18) is uniformly asymptotically stable [L11] in S. Hence

$$\lim_{k \to \infty} f(t) = P^* \qquad (6.19)$$

Also from exercise (6.9) it follows that the normalized vector $\theta^{-\frac{1}{2}}$ [P(k) − f(kθ)] converges in distribution. This inturn implies that E [P(k)] converges [F9]. That is,

$$\lim_{k \to \infty} E\left[p(k)\right] \text{ exists.} \tag{6.20}$$

From corollary (6.1), (6.19) and (6.20) it follows that for any given $\delta' > 0$ and reward parameter a, there exists $0 < \theta^* < 1$ and $b^* < a$ such that for all $0 < b < b^*$ and $0 < \theta < \theta^*$

$$\lim_{k \to \infty} \left| E\left[P(k)\right] - P_{opt} \right| < \delta' \tag{6.21}$$

Substituting (6.21) in (6.2) we obtain

$$\lim_{k \to \infty} \left| \eta(k) - v \right| < \delta' \left| h(dij) \right| \tag{6.22}$$

Where h(dij) is a bounded real valued function of the elements of the matrix. Now for any given $\varepsilon > 0$ choose δ' such that the r. h. s. of (6.22) is less than ε. Q.E.D.

Remark 6.4: If we set $b = 0$ (same as $\theta_2 = 0$) in the L_{R-P}^E algorithm, it reduces to the L_{R-I}^A algorithm discussed in detail in chapter 3. For this choice of the penalty parameter, from theorem 6.4, case 1 (when the game matrix D has no saddle point in pure strategies) we obtain

$$W'(P^*) = W'(P_{opt}) = \begin{bmatrix} 0 & \frac{a}{\Delta}(d_{11}-d_{12})(d_{22}-d_{21}) \\ \\ -\frac{a}{\Delta}(d_{11}=d_{21})(d_{22}-d_{12}) & 0 \end{bmatrix}$$

clearly the eigenvalues of the above matrix are purely imaginary. From this we can expect the mean learning curve to oscillate around the optimal mixed strategy. Such oscillations are reported in [C2][V4]. Also see our example 6 in section 6.4.

The above remark 6.4 and table 6.1 together bring to the limelight the effect of having a <u>small</u> penalty parameter "b". If b = 0 we know that $P^* = P_{opt}$ but the differential equation (6.18) is only stable and <u>not</u> asymptotically stable. In this case we can never expect the solution of the differential equation converge to P^*. On the other hand if b > 0 (but very small compared to the reward parameter a) the solution of the differential equation converges to P^* as t → ∞ even though P^* is only "close" to P_{opt}. Thus, the effect of non-zero value of b is to <u>induce</u> convergence of the solution of (6.18) to a "near" optimal strategies. Once we have the desired properties for the solution of the differential equation, we can analyze the evolution of the mean learning curve by using theorem 6.5. See exercise 6.9 for some of the other properties of the process {P(k)}.

<u>Remark 6.5</u>: Consider the situation when one of the players (say B) uses a fixed (pure or mixed) strategy. This is equivalent to player A playing a game against nature which is the topic of chapters 1 through 4. For definiteness assume player B picks his first and second pure strategies with probability $\tilde{q}_1$ and $1-\tilde{q}_1$ where $0 \leq \tilde{q}_1 \leq 1$. From the analysis presented in chapter 2 it follows that player A, by proper choice of the penalty parameter "b" can make his probability of his gain asymptotically as close to $\max_i \{\tilde{q}_1 d_{i1} + (1-\tilde{q}_1) d_{i2}\}$ as desired. Further it can be easily checked that

$$\max_i \{\tilde{q}_1 d_{i1} + (1-\tilde{q}_1) d_{i2}\} \geq V$$

with equality holding good only when $\tilde{q}_1 = q_{opt}$, that is, only when player B chooses his optimal strategy) where V is the von Neumann value of the game.

6.4 Extensions: -

Virtually all the conclusions obtained by the above analysis hold true when players A and B use different sets of reward and penalty parameters. More specifically let player A use the parameters $\theta_1^A = \theta a_A$, $\theta_2^A = \theta b_A$ and player B use $\theta_1^B = \theta a_B$, $\theta_2^B = \theta b_B$ in place of θ_1 and θ_2 in the algorithm (6.1). Under this set up the following theorem can be easily proved (Exercise 6.11) along the same lines as theorem 6.1.

Theorem 6.6: - For every $0 < a_A < 1$, $0 < a_B < 1$ and $\varepsilon > 0$ there exists $0 < b_A^* < a_A$, $0 < b_B^* < a_B$ and $0 < \theta^* < 1$ such that for all $0 < b_A < b_A^*$ $0 < b_B < b_B^*$ and $0 < \theta < \theta^*$

$$\lim_{k \to \infty} \left| \eta(k) - V \right| < \varepsilon$$

This theorem 6.6 demonstrates the robustness of our method in the sense that both the players can individually choose their reward and penalty parameters in such a way that their asymptotic expected probability of gain can be made arbitrarily close to the Von-Neumann value of the game.

Further by combining the analysis in Chapter 2 and those presented in this chapter, it can be shown that all the conclusions of this chapter remain true when both players use the N_{R-P}^E algorithms as well as when the game matrix is of order M x N (M, N $\geq$ 2). See Exercises 6.7 and 6.8.

6.5 Simulations: A number of simulations, using algorithm (6.1), of games with and without saddlepoint in pure strategies are presented in this section. In all the examples, the mean learning curve is plotted as an average of 25 experiments.

Example 6.1:(continued) The game matrix of example 6.1 is used in this experiment and $E[P(k) \mid p_1(0) = q_1(0) = 1/2]$ are plotted for $\theta = 0.25$ and $\theta = 0.1$ in figures 6.5 and 6.6 respectively. In both cases $a = 0.20$ and $b = 0.01$. See the effect of step-length parameter on the mean-learning curve. Recall that the value of P^* and eigenvalues of $W'(P^*)$ are independent of θ.

Example 6.2 (continued) The mean learning curve for the game matrix of example 6.2 is given in figure 6.7, where $a = 0.20$, $b = 0.001$ and $ = 0.10$.

For a number of other interesting simulations refer [C2] and [V4].

6.6 Comments and Historical Remarks

A number of attempts have been made in the past to analyze the zero sum games without saddle point in pure strategies using absorbing barrier algorithms. Chandrasekaran and Shen [C2] and Viswanathan and Narendra [V4] did a variety of simulations and reported that the value of the game oscillates around the von Neumann value of the game. Similar oscillations are also reported in Tsetlin [T2]. Much of this chapter follows Lakshmivarahan and Narendra [L9]. The linear reward-penalty-L_{R-P}^E algorithm with very small penalty term compared to reward has been recently developed by the author in [L7] in the context of game against nature to attain ε-optimality. See section 5.6 for other related comments.

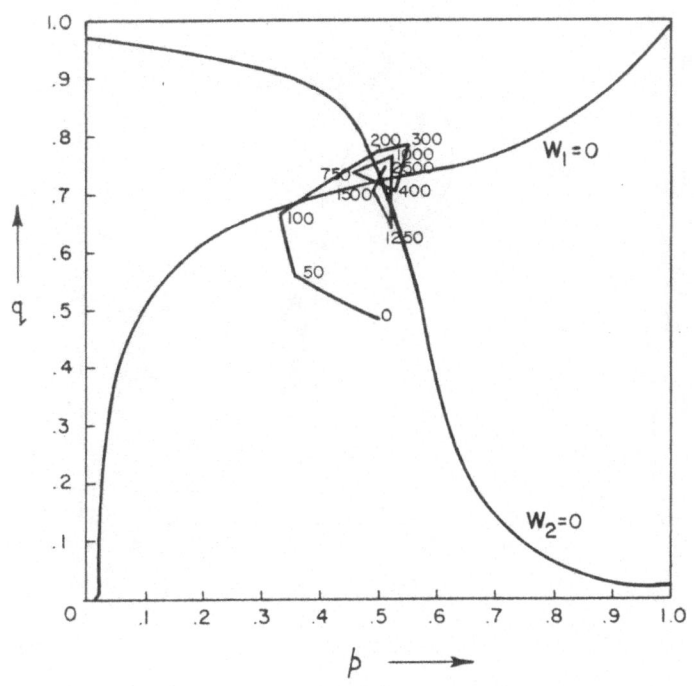

FIGURE 6.5

$$D = \begin{bmatrix} .60 & .20 \\ .35 & .90 \end{bmatrix}$$

a = 0.20   p(2500) = .5180

b = 0.01   q(2500) = .7566

θ = 0.25

AVERAGE OF 25 EXPERIMENTS

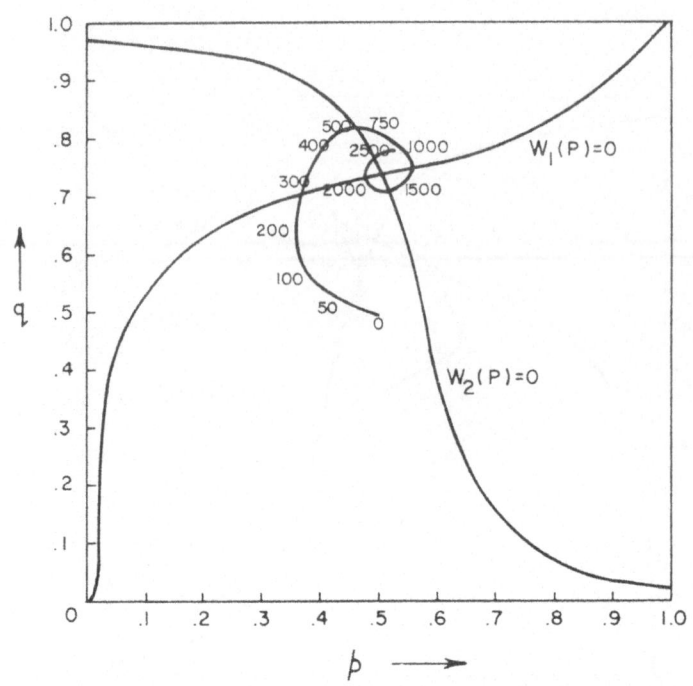

FIGURE 6.6

$$D = \begin{bmatrix} .60 & .20 \\ .35 & .90 \end{bmatrix}$$

a = 0.20   p(2500) = 0.5405

b = 0.01   q(2500) = 0.7801

θ = 0.10

AVERAGE OF 25 EXPERIMENTS

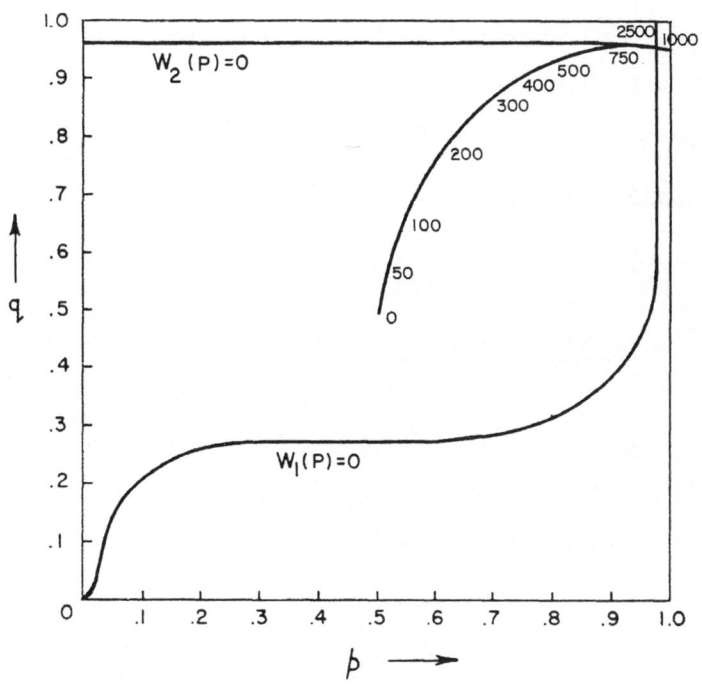

FIGURE 6.7

$$D = \begin{bmatrix} .60 & .80 \\ .35 & .90 \end{bmatrix}$$

a = 0.20   p(2500) = 0.9926

b = 0.001 q(2500) = 0.9873

θ = 0.10

AVERAGE OF 25 EXPERIMENTS

6.7 Appendix

Define

$$h_1(x) \triangleq [(1-x)^2 c_2 - x^2 c_1]$$

$$h_2(x) \triangleq x(1-x)(c_2-c_1) + h_1(x)$$

$$h_3(x) \triangleq x(1-x)(c_2-c_1) + b\, h_1(x)$$

where $x \in [0,1]$, $0 < c_1$, $c_2 < 1$, $0 < b < 1$. We first prove the following:

Lemma 6A.1: There exists unique $x_i^* \in (0,1)$ ($i = 1,2,3$) such that $h_i(x_i^*) = 0$

and

(a) $x_3^* \gtrless x_2^* \gtrless x_1^* \gtrless \frac{1}{2}$ according as $c_2 \gtrless c_1$

(b) x_i^* is an increasing function of (c_2-c_1) for all $i = 1,2,3$

(c) as $b \to 0$, $x_3^* \to 1$ or 0 according as $c_2 > c_1$ or $c_2 < c_1$.

Proof: Let $c_2 - c_1 = \delta$. Note that $\delta \in (-1,1)$. Then

$$h_1(x) = c_1(1-2x) + (1-x)^2 \tag{6A.1}$$

As $h_1(0) = c_2 > 0$ and $h_1(1) = -c_1 < 0$, $h_1(1/2) \gtrless 0$

according as $\delta \gtrless 0$ and $h_1(x)$ is quadratic in x, there exists an unique

$x_1^* \in (0,1)$ and $x_1^* \gtrless \frac{1}{2}$ according as $\delta \gtrless 0$ such that $h_1(x_1^*) = 0$. From (6A.1)

it further follows that λ_1 is an increasing function of δ. Consider

$$h_2(x) = c_1(1-2x) + \delta(1-x) \tag{6A.2}$$

We have $h_2(0) = c_2 > 0$ and $h_2(1) = -c_1 < 0$; $h_2(x_1^*) = x_1^*(1-x_1^*)\delta$ and $h_2(x)$

is linear in x. Thus there exists an unique $x_2^* \in (0,1)$ such that $h_2(x_2^*) = 0$

where $x_2^* \gtrless x_1^*$ according as $\delta > 0$. Also the claim x_2^* is an increasing

function of δ following from (6A.2). Similarly, we have

$$h_3(x) = h_2(x) - (1-b)\, h_1(x)$$

$$h_3(0) = b\, c_2 > 0, \quad h_3(1) = -b\, c_1 < 0$$

and

$$h_3(x_2^*) - (1-b)\, \delta\, x_2^*(1-x_2^*)$$

As $h_s(x)$ is a quadratic in x, there exists a $x_3^* \varepsilon (0,1)$ such that $h_3(x_3^*) = 0$ and $x_3^* \gtrless x_2^*$ according as $\delta \gtrless 0$. Further x_2^* increasing with δ implies x_3^* is also increasing with δ. To prove (c), let $\delta > 0$ and redefine $h_3(.)$ as

$$h_3[b,x] = h_2(x) - (1-b) h_1(x) \qquad (6A.3)$$

Let $x_3^*(b)$ be such that $h_3 [b,x_3^*(b)] = 0$, to emphasize the dependence of x_3^* on b. If $b' < b$, then it can be seen that

$$h_3[b', x_3^*(b)] = h_1(x_3^*(b)) (b'-b) > 0 \qquad (6A.4)$$

From (6A.3) and (6A.4) it follows that $x_3^*(b') > x_3^*(b)$ if $b > b'$. Thus $x_3^*(b)$ increases as b decreases and by making b sufficiently small $x_3^*(b)$ can be made as close to unity as desired. By similar arguments, we can see that $x_3^*(b) \to 0$ as $b \to 0$ when $\delta < 0$. \hfill Q.E.D.

Completion of proof of Theorem 6.3:

We will only give the proof for case 1 (defined by 6.8). Define

$$c_{21}^A(q_1) = c_2^A(q_1) - c_1^B(q_1)$$

$$c_{21}^B(p_1) = c_2^B(p_1) - c_2^B(p_1)$$

It can be checked that

$$c_{21}^A(q_1) = (d_{12} - d_{22}) + q_1 \Delta$$

$$c_{21}^B(p_1) = (d_{22} - d_{21}) - p_1 \Delta$$

where $\Delta = (d_{11} + d_{22}) - (d_{12} + d_{21})$

From (6.8), it follows that $\Delta > 0$ and that $c_{21}^A(q_1)$ $[c_{21}^B(p_1)]$ is an increasing [decreasing] function of q_1 $[p_1]$. From lemma 6A.1, it follows that there exists $p_1(q_1) \varepsilon (0,1)$ and $P(q_1) = (p_1(q_1), q_1)$ such that $W_1(P(q_1))=0$ and $p_1(q_1)$ is an increasing function of q_1. Similarly there exists $q_1(p_1) \varepsilon (0,1)$ and $P(p_1) = (p_1, q_1(p_1))$ such that $W_2(P(p_1)) = 0$ and $q_1(p_1)$

is a decreasing function of p_1. These functions when plotted (refer figure 6.2, 6.3, 6.4) on a unit square have an unique intersection. A similar proof holds for case 2 as well.   Q.E.D.

<u>Proof of Corollary 6.1</u>: Here again we consider only case 1. When b = 0, it can be seen that

$$W(P) = 0 \quad \text{at} \quad P = (1,0)^T, \ (0,1)^T, \ (0,0)^T, \ (1,1)^T$$

and $(p_{opt}, q_{opt})^T$. For b > 0, there exists only one P^* for which $W(P^*)=0$. Using the arguments similar to the one used in proving (c) of lemma 6A.1, we can readily see that by proper choice of b, P^* can be made as close to P_{opt} as desired.   Q.E.D.

6.8   Exercises:

6.1 Show that the algorithm (6.1) together with the game matrix defines
     a random system with complete connection (refer chapter 1 for
     definition).

6.2 If $0 < d_{ij} < 1$ and $0 < \theta_1$, θ_2, < 1, then show the algorithm (6.1) is
     distance diminishing.

6.3 Prove theorem 6.2.

6.4 Compute the elements of the matrix $s(P)$ and prove that it is positive
     definite for all $P \varepsilon S$.

6.5 Verify the properties (P6.3)-(P6.5).

6.6 Complete the proof of theorem 6.5.

6.7 Starting with the general N_{R-P}^E algorithm 2.1, prove all the theorems
     of this chapter.

6.8 Extend the analysis of L_{R-P}^E algorithm 6.1 to the case when players A
     and B have $M(\geq 2)$ and $N(\geq 2)$ pure strategies.

6.9 Show that

$$z(k) = \frac{[P(k)-f(k\theta)]}{\theta}$$

     is such that

$$\lim_{\substack{k\theta \to \infty \\ \theta \to 0}} L(z(k) = N[0,g(\infty)]$$ where the matrix $g(\infty)$ is obtained as the

     unique solution of the linear system

$$W'(P^*) \, g(\infty) + g(\infty) \, [W'(P^*)]^T + s(P^*) = 0,$$

     $s(P)$ is defined in (P6.2), $L(x)$ refers to the distribution of x and
     $N[a,b]$ refers to the normal distribution with mean (vector) a and
     covariance matrix $g(\infty)$.

     Hint: Refer to theorems 2.4 and 2.7.

6.10 Consider the time–varying version of (6.1) obtained by choosing $\theta_1 = \theta(k)a$ and $\theta_2 = \theta(k)b$ where $0 < b < a < 1$ and $\theta(k) \geq 0$ and $\theta(k) \to 0$ as $k \to \infty$

If

$$\sum_{k=0}^{\infty} \theta(k) = \infty \quad \text{and} \quad \sum_{k=0}^{\infty} \theta^2(k) < \infty$$

then show (using the theory of chapter 4, section 4.2) that for every $\varepsilon > 0$, there exists a $0 < b^*$ such that for all $0 < b < b^*$

$$\lim_{k \to \infty} \left| \sum_{i=1}^{2} \sum_{j=1}^{2} p_i(k) \, d_{ij} \, q_j(k) - V \right| < \varepsilon \quad \text{with}$$

probability 1, where V is the Von–Neumann value of the game.

6.11. Prove Theorem 6.6

Chapter 7

Two Person Decentralized Team Problem

With Incomplete Information

7.1. INTRODUCTION:

In this chapter we propose a learning approach to the two person decentralized team problem with incomplete information: Let A and B be the two
persons each with two actions at their disposal. At any instant each person
picks an action (perhaps randomly) and let i be the action chosen by A and j by
B. As a result of their joint actions, both A and B receive the same outcome.
The outcome is in general, a random variable whose distribution depends on the
action pair (i,j). We assume that the outcome is two valued: +1, (unit gain)
and -1 (unit loss) and that neither person has knowledge of the set of actions
available to the other person or the actual action chosen by the other person at
any instant of time or the distribution of the random outcome as a function of
the pair (i,j) of actions. In other words, we consider an interaction between
persons wherein there is no transfer of information of any kind at any stage.
Just based on the action chosen and the random outcome he receives, each person
decides to update the probability distribution over the set of his own actions

using a learning algorithm. In this setup our problem is to find conditions on
the learning algorithm such that asymptotically each person will maximize his
expected outcome or payoff.

When two persons interact, it is conceivable that the outcome they individually receive need not be the same for any given action pair. At the one
end of the scale is the case in which for any action pair A's outcome is equal
in magnitude but opposite in sign to that of B. Such an interaction gives rise
to the well known two person zero sum game model. At the other end of the scale
is the case described in the above paragraph wherein A and B receive the same
outcome. This is called the two person team model. Between these two extremes
is the case where, for any action pair, A and B obtain different outcomes. This
latter situation gives rise to the bimatrix game model which has been extensively
studied under the title Cooperative and Non-Cooperative Games.

The concept of team was first introduced by J. Marschak in 1955 and has
been extensively studied since then [M2] [H5] [H6]. A team is a group or an organization in which all the teammates share a single objective function and in
many ways team decision theory is an extension of the classical Bayesian statistic
decision theory to a group of interdependent decision makers. In the contemporary
literature on team decision theory it is assumed that the organizer of the team
has complete knowledge of (a) the number of members of the team, (b) the set of
all actions available to each member of the team, (c) the teams common payoff
function, (d) the set of all possible states of nature, (e) the information that
each member of the team has about the environment and (f) the (prior) probability
distribution of all relevant quantities - such as the distribution of the states
of nature, the distribution of the information about the environment, etc. A
team is called centralized or decentralized (Chapter 4 in [M2]) depending on
whether all its members do or do not have the same information about the environme

The information of the ith member of the team about the environment may come from two different sources: (1) his own personal observation of the environment and (2) a message communicated to him by other teammates about their knowledge of the environment. The principal problem of the team decision theory may then be stated as follows: "given (a)-(f) and the cost of communication between the members of the team, find the best communication network (for transfer of information among the members of the team) and the best decision rule". The organizer on solving the above problem using one of many known methods [M2] [H5] [H6] would build the optimal communication network and instruct the teammates to implement the optimal decision.

On the contrary, the team problem stated in the first paragraph has no explicit organizer who will precompute the optimal decision rule. Further, each member of the team does not have any other knowledge except his own set of actions. Accordingly, the above team problem is one of decentralized team with incomplete information.

In an attempt to solve our decentralized team problem with incomplete information, one might propose that the teammates should together first empirically estimate all the unknown quantities. Using these estimates, the teammates may jointly compute their optimal actions using well established solution methods [M2]. Such an approach, in analogy with the contemporary literature on adaptive control, may be termed as "indirect method" [N5]. An alternate possibility is to use a "direct method" wherein the teammates learn about the optimal decision using an on-line sequential learning procedure quite analogous to those used in Chapters 5 and 6. In this chapter our interest is in the latter direct approach.

In the team problem under consideration, as A and B have two actions each, their common expected payoff is represented as a 2 x 2 matrix. Given any 2 x 2 matrix, in general, there could be more than one element which is simultaneously the maximum in its row and column. If there is only one element in the payoff matrix with this property we call it a _unimodal_ matrix, otherwise it is called a multimodal matrix.

In this chapter we apply L_{R-P}^{E} and L_{R-I}^{A} algorithms to the decentralized team with incomplete information. These algorithms have been extensively studied in Chapters 2 and 3. Analysis of the team problem with unimodal and multimodal payoff matrix and when the players use the L_{R-P}^{E} algorithm given in Section 7.2. Section 7.3 presents the problem of unimodal payoff matrix and when the players use L_{R-I}^{A} algorithm. A variety of simulation results are given in Section 7.4.

7.2. ANALYSIS OF DECENTRALIZED TEAM PROBLEM - L_{R-P}^{E} ALGORITHM

Consider the two person team problem in which both A and B have two actions each. At any stage k, let A pick his i^{th} action with probability $p_i(k)$ where $p(k) = (P_1(k), p_2(k))$ and $p_1(k) + p_2(k) = 1$, $0 \le p_i(k) \le 1$, $i = 1,2$. Similarly B picks his j^{th} action with probability $q_j(k)$ where $q(k) = (q_1(k), q_2(k)$ and $q_1(k) + q_2(k) = 1, 0 \le q_1(k) \le 1$, $i = 1,2$. Let A and B pick the actions i and j as a sample realization from $p(k)$ and $q(k)$ respectively. For this action pair (i, j), both A and B get a unit gain with probability d_{ij} and unit loss with probability $c_{ij} = 1-d_{ij}$. It is assumed that $0 < d_{ij} < 1$ for i, j = 1,2 and let $D = [d_{ij}]$ be the matrix of d_{ij}. Without loss of generality, we will consider D itself as the payoff matrix.

Let $0 < \theta_i^A < 1$, $0 < \theta_i^B < 1$, $i = 1,2$. At stage k let i and j be the actions actually chosen by A and B respectively. Then p(k+1) is defined as follows:

$$p_i(k+1) = p_i(k) + \theta_1^A \left[1 - p_i(k)\right] \left.\begin{array}{c} \\ \\ \end{array}\right\} \quad \text{if A receives a}$$

$$p_s(k+1) = p_s(k) - \theta_1^A \, p_s(k), s \neq i \quad \text{unit gain}$$

$$p_i(k+1) = p_i(k) - \theta_2^A \, p_i(k) \left.\begin{array}{c} \\ \\ \end{array}\right\} \quad \text{if A receives a}$$

$$p_s(k+1) = p_s(k) + \theta_2^A \, p_i(k), s \neq i \quad \text{unit loss}$$

B uses a similar algorithm with q replacing p, j replacing i and θ_i^B replacing θ_i^A. Also $0 \leq p_s(0) \leq 1$, $0 \leq q_s(0) \leq 1$, $s = 1,2$. The increase (decrease) in the probability $p_i(k)$ when A receives a unit gain (unit loss) is called the reward (penalty) for A. Similarly for B.

Let $\theta_1^A = \theta a_A$, $\theta_2^A = \theta b_A$, $\theta_1^B = \theta a_B$, $\theta_2^B = \theta b_B$. The constant $a_A(a_B)$ is called reward parameter for A(B) and similarly $b_A(b_B)$ is called the penalty parameter for A(B). It is assumed that $0 < b_A < a_A < 1$ and $0 < b_B < a_B < 1$. The constant $0 < \theta \leq 1$ essentially control the step length.

Define

$$\eta(k) = E\left[p(k)\right] \ D \ E\left[q^T(k)\right]$$

where E [.] is the expectation taken over all possible choice of actions by either person and all random outcomes prior to k[th] stage. Let

$$\mathcal{L} = \left\{(m,n) \ \middle| \ \begin{array}{l} d_{mn} \text{ is simultaneously the maximum} \\ \text{in its row and column} \end{array}\right\}$$

If $|\mathcal{L}| = 1$, then D is called unimodal, otherwise it is multimodal. Let $d_{m^*n^*} = \max \ d_{ij}$.

Assumption A: Let D be unimodal and without loss of generality let

$$d_{11} > \max \left\{ d_{12}, d_{22}, d_{21} \right\} \tag{7.3}$$

that is, $\mathcal{I} = \left\{ (1,1) \right\}$ and $d_{m^*n^*} = d_{11}$.

Clearly, under the Assumption A, $\eta(k)$ is maximum only when $p(k) = p_{opt} = (1,0)$ and $q(k) = q_{opt} = (1,0)$.

Our main result is summarized by the following:

Theorem 7.1 Let A and B use the algorithm (7.1). Under Assumption A, for every $\epsilon > 0$, $0 < a_A < 1$ and $0 < a_B < 1$, there exist $0 < b_A^* < a_A$, $0 < b_B^* < a_B$ and $0 < \theta^* < 1$ such that for all $0 < b_A < b_A^*$, $0 < b_B < b_B^*$, and $0 < \theta < \theta^*$

$$\lim_{k \to \infty} \left| \eta(k) - d_{m^*n^*} \right| < \epsilon \tag{7.4}$$

The above theorem, in other words, states that either person can independently choose their reward and penalty parameters and asymptotically achieve a payoff that is as close to the maximum as desired. It will become apparent in the course of the proof that b_A^*, b_B^*, and θ^* depend only on ϵ and not on the initial values of the vectors $p(0)$ and $q(0)$.

Analysis:

Let $S = [0,1] \times [0,1]$, $I = \{1,2\}$, $E = \{\text{unit gain, unit loss}\}$, $E_1 = I \times I \times E$. Define $p(k) = (p_1(k), q_1(k))$. Clearly $\{P(K)\}$ $k \geq 0$ is a Markov process with stationary transition function with S as its state space. E is called the (finite) event space. If $s = (p_1, q_1)$ S, $e^1 = (i, j, e)$, then algorithm (7.1) defines a mapping $T: S \times E_1 \to S$ where

$$s(k+1) = T [s(k), e^1(k)].$$

The event probability distribution is given below:

$$\text{Prob } [e^1=(i, j, e) \mid s = (p_1, q_1)] = \mathcal{K}[e^1, s]$$

$$\mathcal{K}[e^1, s] = p_i \, q_j \, d_{ij} \text{ if } e = \text{unit gain}$$

$$= p_i \, q_j \, c_{ij} \text{ if } e = \text{unit loss}$$

where recall $p_2 = 1 - p_1$ and $q_2 = 1 - q_1$.

<u>Remark 7.1</u>: The quadruple $(S, E, T, \mathcal{K})$ is known as the random system with complete connection. Further, since $0 < \theta_i^A < 1$ and $0 < \theta_i^B < 1$, $i = 1,2$ and $0 < d_{ij} < 1$ it follows from algorithm (7.1) that $s(k+1) \neq s(k)$ with probability one. In other words, all the states of the Markov process $\{P(k)\}$ $k \geq 0$ are non-absorbing (Chapters 2 and 5).

Let $\delta P(k) = (\delta p_1(k), \delta q_1(k))$ where $\delta z(k) = z(k+1) - z(k)$.

It can be shown that

(P7.1) $\qquad E[\delta P(k) \mid P(k) = P = (p_1, q_1)] = \theta W(P)$

where

$$W(P) \quad = \quad (W_1(P), W_2(P))$$

$$W_i(P) \quad = \quad W_i^R(P) + W_i^P(P), \; i = 1,2$$

$$W_1^R(P) \quad = \quad a_A \, p_1 \, (1-p_1) \, [c_2^A(q_1) - c_1^A(q_1)]$$

$$W_2^R(P) \quad = \quad a_B \, q_1 \, (1 - q_1) \, [c_2^B(p_1) - c_1^B(p_1)]$$

$$W_1^P(P) \quad = \quad b_A \, [(1 - p_1)^2 \, c_2^A(q_1) - p_1^2 \, c_1^A(q_1)]$$

$$W_2^P(P) \quad = \quad b_B \, [(1 - q_1)^2 \, c_2^B(p_1) - q_1^2 \, c_1^B(p_1)]$$

and

$$c_i^A(q_1) = c_{i1}q_1 + c_{i2}(1 - q_1)$$

$$c_i^B(p_1) = c_{1i}p_1 + c_{2i}(1 - p_1)$$

Similarly

(P7.2) $\qquad E\left\{[\delta P(k) - W(P)] \, [P(k) - W(P)]^T \mid P(k) = P = (p_1, q_1)\right\} = \theta^2 s(P)$

where the elements of s(p) are easily computed. Since all the states of the process {P(k)} k≥0 are non-absorbing s (P) is positive definite uniformly for all P ε S. Further

(P7.3) $E \ \{ \ |P(k)|^3 \qquad P(k) = P = (p_1, q_1)\} = 0 \ (\theta^3)$

uniformly for all p ε S and |.| refers to the Euclidean norm on S.

(P7.4)   W(P) has bounded Lipschitz derivative is S.

(P7.5)   s(P) is Lipschitz in S.

<u>Remark 7.2</u>: As is evident from (P7.1) the function W(P) refers to mean normalized (by θ) step length (also known as the <u>drift</u>) function of the Markov process P(k). E [P(k) | P(0)] is known as the "<u>mean learning curve</u>". s(P) refers to the one step covariance matrix. Since s(P) is positive definite for all **p ∈** S, the asymptotic behavior of the Markov process crucially depends on the properties (such as zeros, etc.) of W(P). Properties (P7.3)-(P7.5) are easily verified and we omit the details.

The following theorems summarize further properties of W(P).

<u>Theorem 7.2</u>: <u>For every</u> ε > 0, <u>there exists</u> $0 < b_A^* < a_A$, $o < b_B^* < a_B$ <u>such that for</u> <u>all</u> $0 < b_A < b_A^*$ <u>and</u> $0 < b_B < b_B^*$ <u>there exists a unique</u> $p^* = (p_1^*, q_1^*)$ <u>such that</u>
$$W(P^*) = 0$$

and

$$|P^* - P_{opt}| < ε$$

where $P_{opt} = (1,1)$ <u>under Assumption</u> A.

<u>Proof</u>: The basic idea of the proof of this Theorem is developed in lemma 6A.1 (Section 6.7, Chapter 6).

Define

$$c_{21}^A (q_1) = c_2^A (q_1) - c_1^A (q_1)$$

$$c_{21}^B (p_1) = c_2^B (p_1) - c_1^B (p_1).$$

It can be seen that

$$c_{21}^A (q_1) = (d_{12} - d_{22}) + q_1 \cdot \triangle$$

$$c_{21}^B (P_1) = (d_{21} - d_{22}) + p_1 \cdot \triangle$$

where

$$\triangle = (d_{11} + d_{22}) = (d_{12} + d_{21})$$

Assumption A gives rise to the following three cases:

<u>Case A:</u>

$$d_{21} \leq d_{22} < d_{12} < d_{11} \text{ or } d_{21} < d_{22} \leq d_{12} < d_{11}$$

<u>Case B:</u>

$$d_{12} \leq d_{22} < d_{21} < d_{11} \text{ or } d_{12} < d_{22} \leq d_{21} < d_{11}$$

<u>Case C:</u>

$$d_{11} > \max \{d_{12}, d_{21}\} \text{ and } d_{22} < \min \{d_{12}, d_{21}\}$$

Consider first the Case A. In this case, $\triangle > 0$, $c_{21}^A (q_1) > 0$ for $q_1 \in [0,1]$ and $c_{21}^A (q_1)$ is the monotically increasing function of q_1. For each fixed q_1 and $o < b_A < a_A$, it can be seen that $W_1(P)$ has all the properties of the function $h_3(x)$ of lemma 6A.1. Hence, by the same lemma, for each fixed q_1, there exists a unique $p_1(q_1)$ such that $W_1(P(q_1)) = 0$, $P(q_1) = (p_1(q_1), q_1)$ $p_1(q_1) > 1/2$ for all q_1 and $p_1(q_1)$ is a monotonically increasing function of q_1. Further, by making b_A small, one can make $p_1(q_1)$ as close to unity as desired.

Consider now $W_2(P)$. It can be seen that

$$
c_{21}^B(p_1) \quad
\begin{cases}
< 0 & \text{if } p_1 < d* \\
= 0 & \text{if } p_1 = d* \\
> 0 & \text{if } p_1 > d*
\end{cases}
$$

where

$$
d* = \frac{d_{22} - d_{11}}{\triangle}
$$

We obtain from lemma 6A.1 that for each fixed $p_1 \varepsilon$ [0,1] and for $0 < b_B < a_B$, there exists a unique $p_1(q_1)$ such that $W_2(P(p_1)) = 0$, $P(p_1) = (p_1 q_1(p_1))$,

$$
q_1(p_1) \quad
\begin{cases}
< 1/2 & \text{for } p_1 < d* \\
= 1/2 & \text{for } p_1 = d* \\
> 1/2 & \text{for } p_1 > d*
\end{cases}
$$

and that $q_1(p_1)$ is the monotonically increasing function of p_1. Also by making b_B small, $q_1(p_1)$ can be made close to 0 for all $p_1 < d*$ and close to 1 for all $p_1 > d*$.

It can be seen that the functions $p_1(q_1)$ and $q_1(p_1)$ defined above when plotted on unit square has unique intersection $P* = (p*, q*)$ such that $W_i(P*) = 0$, $i = 1,2$. It also follows from the above discussion that for every $\varepsilon > 0$ there exists $0 < b_A^* < a_A$ and $0 < b_B^* < a_B$ such that for all $0 < b_A < b_A^*$ and $0 < b_B < b_B^*$ we can make $P*$ lie in an ε - neighborhood of P_{opt} = (1 This concludes the proof of Theorem 7.2 for the Case A. The proof for the other two cases follows along similar lines.

$$Q.E.D.$$

All the other ways in which D is unimodal can be obtained from the above three cases by interchanging the rows and/or columns of D.

Define

$$
W'(P) = \begin{bmatrix} \dfrac{\delta\,W_1\,(P)}{\delta\,p_1} & \dfrac{\delta\,W_1(P)}{\delta\,q_1} \\[4em] \dfrac{\delta\,W_2\,(P)}{\delta\,p_1} & \dfrac{\delta\,W_2(P)}{\delta\,q_1} \end{bmatrix}
$$

where

$$
\frac{\delta\,W_1\,(P)}{\delta\,p_1} = a_A \left[1-2p_1\right]\,[c_2^A\,(q_1) - c_1^A\,(q_1)]
$$

$$
- 2b_A\,[(1-p_1)\,c_2^A\,(q_1) + p_1 c_1^A\,(q_1)]
$$

$$
\frac{\delta\,W_1\,(P)}{\delta\,q_1} = a_A\,p_1(1-p_1)\,\triangle + b_A\,[(1-p_1)^2\,(d_{22} - d_{21}) - p_1^2\,(d_{12} - d_{11})]
$$

$$
\frac{\delta\,W_2(P)}{\delta\,p_1} = a_B\,q_1(1-q_1)\triangle + b_B\,[(1-q_1)^2\,(d_{22} - d_{12}) - q_1^2\,(d_{21} - d_{11})]
$$

and

$$
\frac{\delta\,W_2\,(P)}{\delta\,q_1} = a_B\,[1-2p_1]\,[c_2^B(p_1) = c_1^B(p_1)]
$$

$$
- 2b_B\,[(1-q_1)\,c_2^B(p_1) + q_1\,c_1^B(p_1)]
$$

where

$$\triangle = (d_{11} + d_{22}) - (d_{21} + d_{12})$$

Theorem 7.3   Under Assumption A, $W'(P*)$ is negative definite.

Proof:

In view of an Assumption A, $P* \approx P_{opt} = (1,1)$ and

$$W(P*) \approx \begin{bmatrix} a_A (d_{11} - d_{21}) - 2b_A (1-d_{11}), & b_A (d_{11} - d_{12}) \\ \\ \\ b_B (d_{11} - d_{21}), & -a_B (d_{11} - d_{12}) - 2b_B (1-d_{11}) \end{bmatrix}$$

u = Trace of $W'(P*)$ = $-a_A (d_{11} - d_{21}) + a_B (d_{11} - d_{12}) + 2(1-d_{11}) (b_A + b_B)$

     $< \quad 0$

v = Determinant of $W'(P*)$

     $= (a_A a_B - b_A b_B) (d_{11} - d_{12}) (d_{11} - d_{21})$

     $+ \quad 2(1-d_{11}) a_A b_B (d_{11} - d_{21}) + a_B b_A (d_{11} - d_{12}) + b_A b_B (1-d_{11})^2$

     $> \quad 0$

Hence $W'(P*)$ is negative definite [B6] that is, the eigenvalues of $W'(P*)$ have negative real part.   Q.E.D.

With these preliminaries we now proceed to prove Theorem 7.1.

Proof of Theorem 7.1. Combining (P7.1)-(P7.5), Theorem 7.2 and the fact that $W'(P*)$ is negative definite, it follows that the process $\{P(k)\}$, $k \geq 0$ satisfies all the conditions of Theorem 2.7 (Chapter 2). Hence, from the conclusions of that theorem, we obtain

$$E \ [P(k) \ | \ P(0)] \quad = f(k) + 0 \ (\theta) \tag{7.5}$$

for all $k \geq 0$, uniformly for all $P(0) \ \epsilon \ S$ where

$$f'(t) = W(f(t)) \tag{7.6}$$

and $f(0) = P(0)$. From the properties of $W(P)$ given above it follows that the

differential equation (7.6) is uniformly asymptotically stable in S [L11] and that

$$f(t) \to p* \text{ as } t \to \infty \qquad (7.7)$$

Also from the same Theorem 2.7 we obtain that the normalized random vector

$\xi(k) = \theta^{-1/2}$ [P(K) - f(k\theta)] converges in distribution. This in turn implies

that $E [\xi(K)]$ converges and hence

$$\lim_{k \to \infty} E [P(k)] \quad \text{exists}$$

$$(7.8)$$

From Theorem 7.2, and (7.5), (7.7), (7.8) it follows that for any given $\delta > 0$,

there exists a $0 < \theta^* < 1$ such that for all $0 < \theta < \theta^*$

$$\lim_{k \to \infty} \left| E [P(k)] - P_{opt} \right| < \delta \qquad (7.9)$$

Substituting (7.6) in (7.1) we see that

$$\lim_{k \to \infty} \left| \eta(k) \quad -d_{m}* _{n}* \right| < \delta \left| h [d_{ij}] \right| \qquad (7.10)$$

Where $h [.]$ is a bounded function of the elements of D. Now given any $\varepsilon > 0$,

we can choose δ such that the r.h.s. of (7.10) is less than ε and this concludes

the proof of Theorem 7.1.   Q.E.D.

Remark 7.3:   The Theorems 7.1 and 7.2 above together bring to the limelight the

role of the reward (a_A, a_B), penalty (b_A, b_B), step length (θ) parameters and

the importance of the function $W(P)$.   From Theorem 7.1, it is clear that by

proper choice of the reward and penalty parameters we can place the (unique) zero

of $W(P)$ arbitrarily close to the optimal value.   From (7.5) and (7.6) it follows

that the solution of the differential equation (7.6) governed by $W(P)$ well

approximates (to the order of θ) the evolution of the mean learning curve.   Further,

there is an inherent conflict in the choice of θ namely, the larger the value of

θ, the larger will be the speed of convergence but at the cost of increased error

in the approximation given by (7.5) and vice versa.   Thus the best choice for the step length parameter needs further investigation.

Remark 7.4: Theorem 7.1 further demonstrates the robustness of our method in the sense that both the members of the team can individually choose their reward and penalty parameters in such a way that their asymptotic payoff is as close to the maximum as desired.

Remark 7.5: A careful scrutiny of the proof of Theorem 1 reveals that the unimodality of D is used only to ensure the existence and uniqueness of the zero of $W(P)$ and negative definiteness of $W'(P)$. In fact, the theorem is applicable to any payoff matrix—unimodal or otherwise, that will ensure the conclusions of Theorems 7.2 and 7.3. This fact will be verified in a number of multimodal cases discussed in detail below.

Example 7.1: Consider a unimodal matrix

$$
D = \begin{bmatrix} 0.9 & 0.6 \\ \\ 0.7 & 0.4 \end{bmatrix}
$$

The plots of $W_i(P) = 0$, $i = 1,2$ for different values of the reward and penalty parameters are given in figures (7.1) and (7.2).

Now let us go on to analyze the two person team problem when the payoff matrix is multimodal, that is D having two, three and four modes. In the following wherever applicable we only find the conditions for the existence of unique P* such that $W(P*) = 0$ and that $W'(P*)$ is negative definite. Once the above propertie of $W(P)$ are ensured, in view of Remark 7.5, all the convergence results of Theorem 7.1 are applicable and to save space we will not repeat the proof.

a) Matrix D has four modes: This is perhaps the simplest of the team problems with multimodes. Let $\mathcal{I} = \{(1,1), (2,1), (1,2), (2,2)\}$. This can happen only when

$$
d_{ij} = d \text{ and } 0 < d < 1.
$$

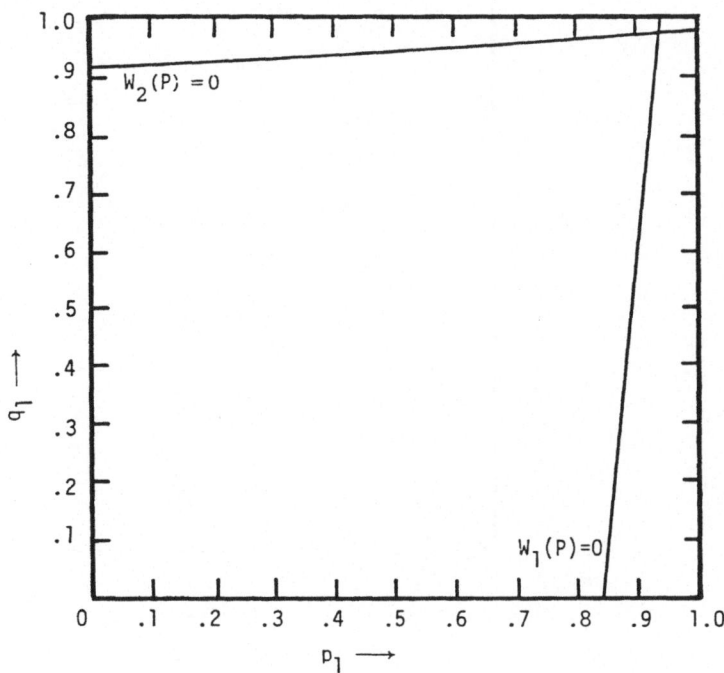

FIGURE 7.1

$$a_A = 0.1 \qquad b_A = 0.01$$

$$a_B = 0.1 \qquad b_B = 0.01 \qquad\qquad D = \begin{bmatrix} 0.9 & 0.6 \\ 0.7 & 0.4 \end{bmatrix}$$

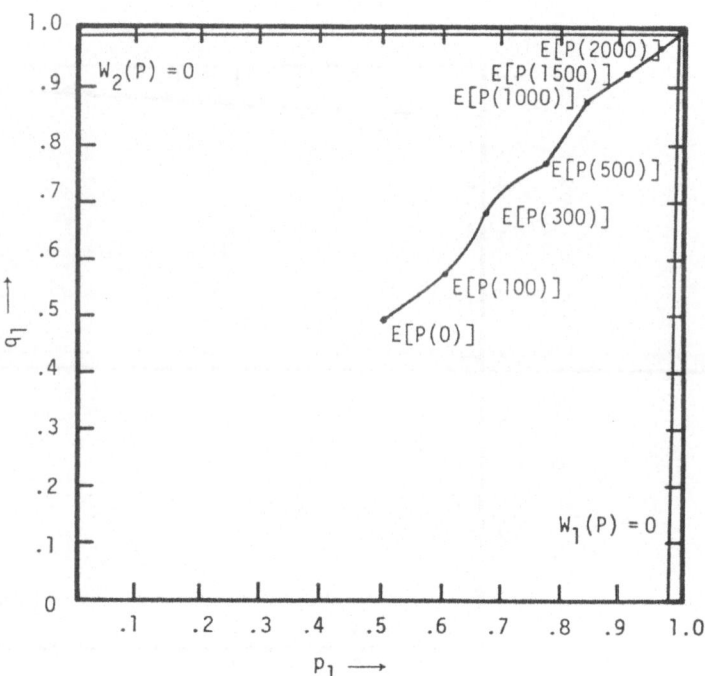

FIGURE 7.2

$a_A = 0.1$ $b_A = 0.001$

$a_B = 0.1$ $b_B = 0.001$

$$D = \begin{bmatrix} 0.9 & 0.6 \\ 0.7 & 0.4 \end{bmatrix}$$

It can be shown that

$$W_i^R(P) = 0 \qquad i = 1,2$$

$$W_1^P(P) = b_A \, (1-d) \, (1-2p_1)$$

and

$$W_2(P) = b_B \, (1-d) \, (1-2q_1)$$

Thus, there exists a unique $P* = (1/2, 1/2)$ such that $W_i(P*) = 0$, $i = 1,2$ and

$$W'(P*) = \begin{bmatrix} -2b_A(1-d) & 0 \\ & \\ 0 & -2b_B(1-d) \end{bmatrix}$$

which is negative definite. Thus all the conclusions of Theorem 7.1 hold. In this case, since any value of P is optimal (in the sense of maximizing $\eta(k)$), $p* = (1/2, 1/2)$, in particular, is optimal. In fact, this in the only situation in which the members of the team using the algorithm (7.1) will ever receive the optimal payoff.

b) <u>Matrix D has three modes:</u>   Let $\mathcal{I} = \{(1,1), (1,2), (2,1)\}$.

This can happen only when

$$d = d_{11} = d_{12} = d_{21} > d_{22} = d'$$

Thus $\Delta = (d' - d) < 0$. It can be seen that for each q_1, there exists unique $P(q_1) = (p_1(q_1), q_1)$ such that

$$W_1(P(q_1)) = 0$$

where

$$p_1(q_1) \begin{cases} > & 1/2 \quad \text{for } q_1 \, \epsilon [0,1] \\ & \\ = & 1/2 \quad \text{for } q_1 = 1 \end{cases}$$

and $p_1(q_1)$ is a monotonically decreasing function of q_1. Similarly, for each p_1 fixed, there exists unique $P(p_1) = (p_1, q_1(p_1))$ such that

$$W_2(P(p_1)) = 0$$

where

$$q_1(p_1) \begin{cases} > & 1/2 \quad \text{for } p_1 \in [0,1) \\ = & 1/2 \quad \text{for } p_1 = 1 \end{cases}$$

and $q_1(p_1)$ is monotonically decreasing. The graphs of $p_1(q_1)$ and $q_1(p_1)$ when plotted on unit square has a unique intersection $P^* = (p_1^*, q_1^*)$. Further by choosing a_A and a_B to be small we can make P^* as close to $(1,1)$ which is optimal and

$$W'(P^*) \approx \begin{bmatrix} -2b_A (1-d) & 0 \\ & \\ & \\ 0 & -2b_B(1-d) \end{bmatrix}$$

is clearly negative definite. Thus all the convergence results of Theorem (7.2) is applicable to this case. All the other ways in which the Matrix D can have three modes can be obtained from the above by interchanging the rows/or columns of D.

Example 7.2: Consider a matrix with three modes

$$D = \begin{bmatrix} 0.6 & 0.6 \\ & \\ & \\ 0.6 & 0.3 \end{bmatrix}$$

Refer to Figures (7.3) and (7.4) for plots of $W_i(P)$, $i = 1,2$ for different values of reward and penalty parameters.

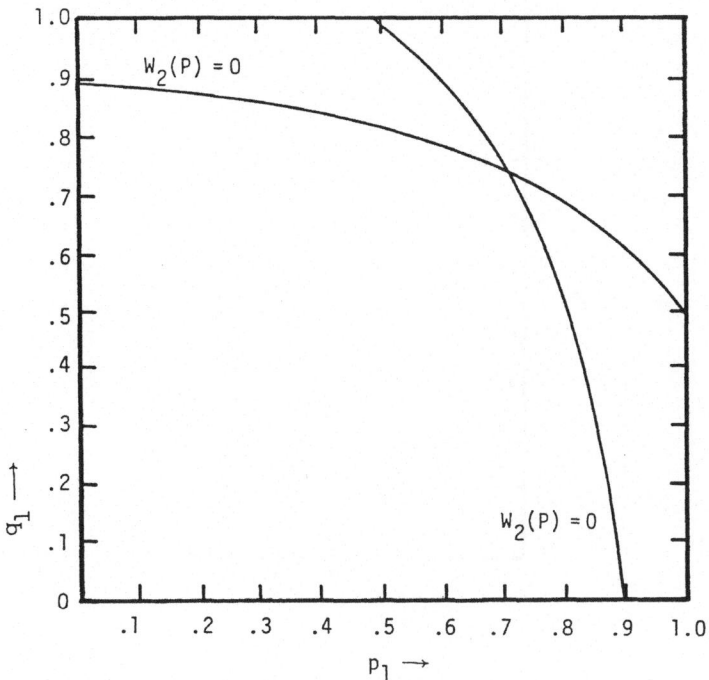

FIGURE 7.3

$a_A = 0.1$ $b_A = 0.01$

$a_B = 0.1$ $b_B = 0.01$

$$D = \begin{bmatrix} 0.6 & 0.6 \\ 0.6 & 0.3 \end{bmatrix}$$

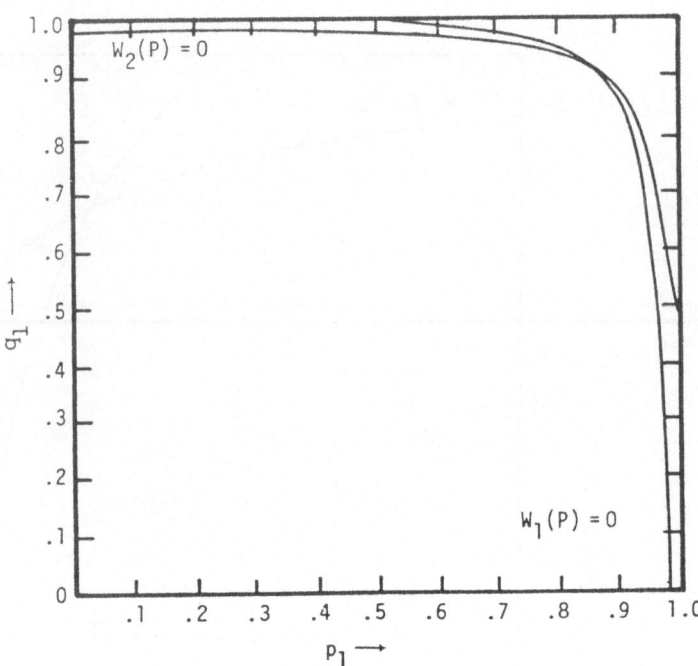

FIGURE 7.4

$a_A = 0.1$ $b_A = 0.001$

$a_B = 0.1$ $b_B = 0.001$

$$D = \begin{bmatrix} 0.6 & 0.6 \\ 0.6 & 0.3 \end{bmatrix}$$

c) Matrix D has two modes: We shall distinguish two subcases

(1)   Let $\mathfrak{X} = \{(1,1), \quad (1,2)\}$ where

$$d = d_{11} = d_{12} > \max \{d_{21}, d_{22}\},$$

that is, the two modes occur along the first row of D. Under these conditions it can be shown that for each q_1 there exists $P(q_1) = (p_1(q_1), q_1)$ such that

$$W_1 (P(q_1)) = 0$$

and $p_1 (q_1) > 1/2$ for all $q_1 \epsilon [0,1]$.

By choosing a_A small enough we can, in fact, make $p_1(q_1)$ as close to unity as desired uniformly for all $q_1 \epsilon [0,1]$.

Assume for definite $d_{21} = d_{22} = d'$. Then $c_2^B(p_1) = c_1^B(p_1)$ for all $p_1 \epsilon[0,1]$ and that there exists unique $P(p_1) = (p_1, q_1 (p_1))$ such that

$$W_2(P(q_1)) = 0$$

and

$q_1(p_1) = 1/2$ for all $p_1 \epsilon [0,1]$. The plots of $p_1(q_1)$ and $q_1(p_1)$ have a unique intersection $P* = (p_1^*, 1/2)$. Further, for any given $\delta > 0$, there exists a $b_A^* > 0$ such that for all $0 < b_A < b_A^*$, $p_1^* > 1 -\delta$. It can be shown that

$$W'(P*) \simeq \begin{bmatrix} -b_A (d-d') - a_A (1-d) & 0 \\ & \\ & \\ 0 & -2a_B (1-d) \end{bmatrix}$$

and hence is negative definite. Now it follows from Theorem 1 that there exists $\theta* > 0$ such that for all $0 < \theta < \theta*$ and $0 < b_A < b_A^*$ we have

$$\lim_{k \to \infty} \eta(k) \leq d + \delta \quad [d + (1/2) (d + d^1)]$$

For any given $\varepsilon > 0$, by proper choice of δ, we obtain the inequality (7.4).

Similar analysis holds good for other cases when $d_{21} \neq d_{22}$.

2) Let $I = \{(1,1), (2,2)\}$ where

$$d_{11} > \max \{d_{12}, d_{21}\} \tag{7.11}$$

$$d_{22} > \max \{d_{12}, d_{21}\}$$

That is, the two modes of D are occurring along the main diagonal of the Matrix D. For definiteness, assume further that $d_{22} > d_{11}$. Under these conditions $\eta(k)$ is maximum only when $P(k) = P_{opt} = (0,0)$. Also $\Delta > 0$ and $c_{21}^A(q_1)$ is an increasing function of q_1 with $c_{21}^A(0) < 0$ and $c_{21}^A(1) > 0$. This implies that for each q_1 there exists $P(q_1) = (p_1(q_1), q_1)$ such that $W_1(P(q_1)) = 0$ and $p_1(0) < 1/2$ and $p_1(1) > 1/2$. Similarly $c_{21}^B(p_1)$ is an increasing function with $c_{21}^B(0) < 0$, $c_{21}^B(1) > 0$ and there exists $P(p_1) = (p_1, q_1(p_1))$ such that $W_2(P(p_1)) = 0$ with $p_1(0) < 1/2$ and $p_1(1) > 1/2$.

Example 7.3: The plots of $p_1(q_1)$ and $q_1(p_1)$ for various values of reward and penalty parameters are given in Figures (7.5), (7.6) and (7.7) when $a_A = a_B = a$ and $b_A = b_B = b$ and

$$D = \begin{bmatrix} 0.6 & 0.2 \\ & \\ & \\ 0.35 & 0.9 \end{bmatrix}$$

It is seen from Example 7.3 that there exists an $b^* > 0$ for such that for all $0 < b^* < b < a$ there exists a unique intersection $P^* = (p_1^*, q_1^*)$. For $0 < b < b^*$, $p_1(q_1)$ and $q_1(p_1)$ intersect at three points. This is, however, to be expected since $W_i(P)$ are continuous in and when $b = 0$.

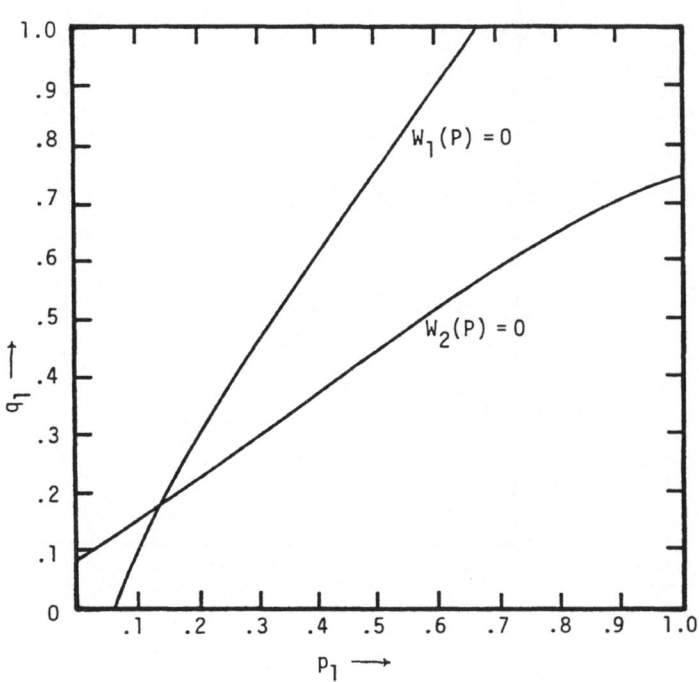

FIGURE 7.5

$a_A = 0.1$ $b_A = 0.05$

$a_B = 0.1$ $b_B = 0.05$

$$D = \begin{bmatrix} 0.6 & 0.2 \\ 0.35 & 0.9 \end{bmatrix}$$

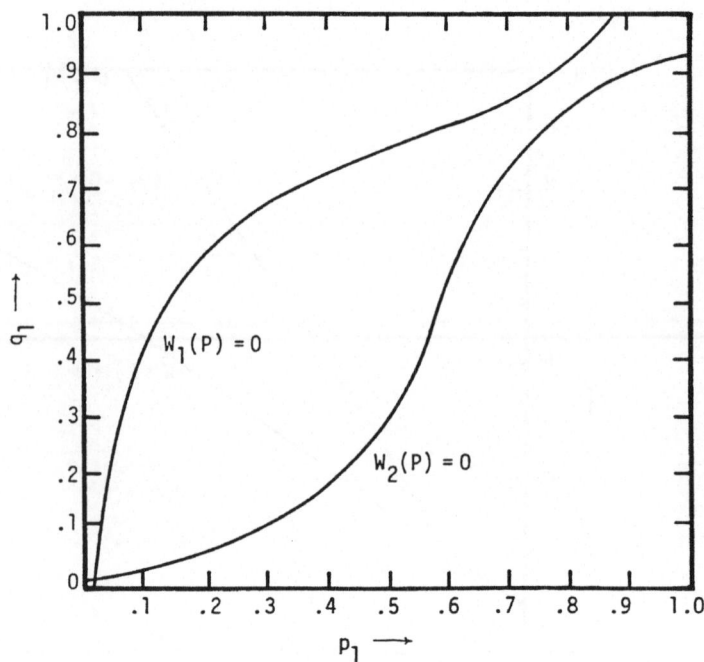

FIGURE 7.6

$a_A = 0.1$ $b_A = 0.01$

$a_B = 0.1$ $b_B = 0.01$

$$D = \begin{bmatrix} 0.6 & 0.2 \\ 0.35 & 0.9 \end{bmatrix}$$

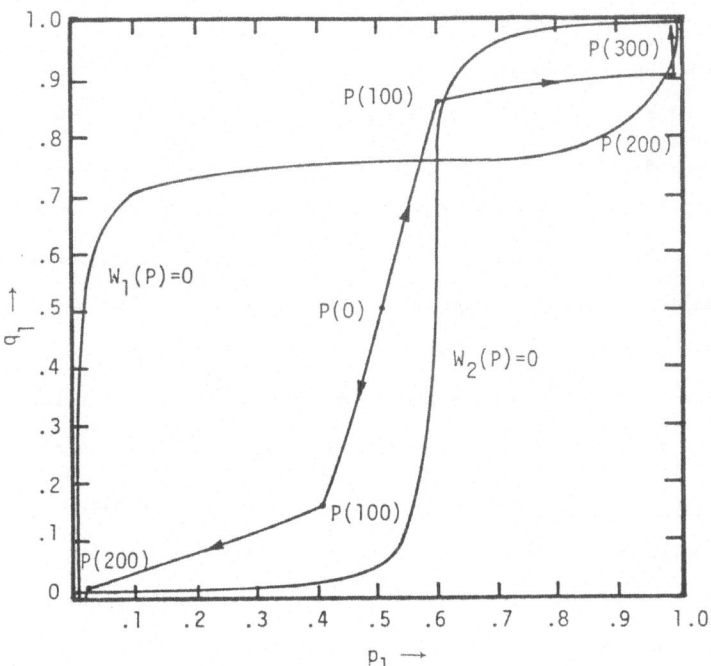

FIGURE 7.7

$a_A = 0.1$ $b_A = 0.001$

$a_B = 0.1$ $b_B = 0.001$

$$D = \begin{bmatrix} 0.6 & 0.2 \\ 0.35 & 0.9 \end{bmatrix}$$

$W_1(P) = 0$ all along $p_1 = 0,1$ and $q_1 = (d_{22} - d_{12}) \big/ \Delta$

$W_2(P) = 0$ all along $q_1 = 0,1$ and $p_1 = (d_{22} - d_{21}) \big/ \Delta$

As a consequence we have the following:

(i) Limiting the penalty parameter b by a lower bound b^* ensures the uniqueness of the solution of $W_i(P) = 0, i = 1,2$. If P* is the unique solution to this set of equations, it can be shown that $W'(P*)$ is negative definite for all $b^* < b < a$. Thus, all the conditions of the Thorem 2.7 of Chapter 2 hold and all the conclusion of this Theorem also carry over to this case. But unfortunately any lower bound on b implicitly defines an upper bound on the limiting value of $\eta(k)$ since P* is bounded away from $P_{opt} = (0,0)$.

(ii) For $0 < b < b^*$ we have seen that there are multiple solutions to $W^i(P) = 0$, Referring to Figure (7.7) denote the solution near (0,0) as $P^a = (p_1^a, q_1^q)$, near $\left(\dfrac{d_{22} - d_{21}}{\Delta} \quad \dfrac{d_{22} - d_{12}}{\Delta}\right)$ as $P^b = (p_1^b, q_1^b)$ and the one near (1,1) as $P^c = (p_1^c, q_1^c)$. It can be shown that $W'(P^a)$ and $W'(P^c)$ are negative definite but $W'(P^b)$ is not. Further the multiple solutions of $W(P) = 0$ prohibits the application of Theorem (2.7) of Chapter 2 and hence of Theorem (7.1) to this case. Refer Section (7.4) for further comments.

7.3 ANALYSIS OF DECENTRALIZED TEAM $- L_{R-I}^A$ ALGORITHM:

In this section we analyze the behaviour of two person decentralized team problem when both the members use the L_{R-I}^A algorithm with the same parameters (that is, with $\theta_2^A = \theta_2^B = 0$ and $\theta_1^A = \theta_1^B = \theta_1 = \theta a$ in 7.1).

Define:

$$\eta_1 \ (k) = E \ [p(k) \ D \ q^T \ (k)] \qquad\qquad (7.12)$$

where E [.] is the expectation taken over all possible choice of actions by either persons and all possible random outcomes where D is a 2 x 2 unimodal matrix with

$d_{m^* n^*} = d_{11}$. Both the members of the team try to maximize $n_1(k)$. Recall $0 < d_{ij} < 1$ for $i,j = 1,2$ and $0 < \theta_1 < 1$. To simplify our presentation we further assume[1] that

$$d_{22} < \min \{d_{12}, d_{21}\} \tag{7.13}$$

Combining (7.13) with the definition of unimodality we obtain

$$d_1^A(k) > d_2^A(k) \text{ and } d_1^B(k) > d_2^B(k) \tag{7.14}$$

Uniformly for all $k \geq 0$, where

$$d_i(k) = \sum_{j=1}^{2} d_{ij} \, q_j(k)$$

and

$$d_j^B(k) = \sum_{i=1}^{2} d_{ij} \, p_i(k) \tag{7.15}$$

Clearly, $\{p(k)\}$ and $\{q(k)\}$ are two non-stationary Markov processes on the unit simplex S_2 of dimension two. Recall $V_2 = \{(0,1)^T, (1,0)^T\}$ is the set of all vertices of S_2.

An alert reader should have recognized the parallelism between the development of this section thus far and that of Section 5.4. All the conclusions listed below follow from Theorem 5.6 (Exercise 7.5).

---

[1] The matrix of Example (7.1) satisfies this condition. However, they are unimodal matrices not satisfying (7.13) such as

$$D = \begin{bmatrix} .9 & .6 \\ .7 & .65 \end{bmatrix}$$

(7C.1)
$$\lim_{k \to \infty} p(k) \ \epsilon \ V_2$$

$$\left.\begin{array}{c} \\ \\ \\ \\ \\ \end{array}\right\}$$ with probability 1   (7.16)

$$\lim_{k \to \infty} q(k) \ \epsilon \ V_2$$

(7C.2) For every $\epsilon_1 > 0$, there exists an $0 < a^* < 1$ such that for all $0 < a < a^*$

$$\text{Prob} \left[\lim_{k \to \infty} p(k) = (1,0)^T \ \Big| \ p(0) \right] > 1 - \epsilon_1 \qquad (7.17)$$

and

$$\text{Prob} \left[\lim_{k \to \infty} q(K) = (1,0)^T \ \Big| \ q(0) \right] > 1 - \epsilon_1 \qquad (7.18)$$

where $0 < p_s(0) < 1$, $o < q_s(0) < 1, s = 1,2$

(7C.3) Combining (7.12), (7.17), and (7.18) it follows that for every $\epsilon > 0$ by proper choice of the parameter a we obtain

$$\lim_{k \to \infty} | \ n_1(k) - d_{11}| \ < \epsilon \qquad (7.19)$$

In other words, by proper choice of the parameters in the L_{R-I}^A algorithm both the persons will receive a payoff as close to the maximum as desired.

Remark 7.6: Suppose (7.13) is not true. Unimodality of D still demands that d_{22} satisfy either of the following: (a) $d_{21} < d_{22} < d_{12}$ or (b) $d_{12} < d_{22} < d_{21}$. If (a) is true, then (7.17) is ture. That is, person A, independent of what B does, will pick his action 1 asymptotically with a probability as close to unity as desired. Now conditioned on such a behavior of A, it can be shown that B will also pick his action 1 with a probability as close to unity as desired. In other words, A will "lead" and B will "follow". Similar analysis will hold good when (b) is true.

7.4. SIMULATIONS:

Example 7.1 (continued): The algorithm (7.1) was simulated with $a_A = a_B$
= 0.1 and $b_A = b_B$ = 0.001 and $\theta = 0.5$ and the mean learning curve (computed as
the numerical average over 50 sample runs) is plotted in Figure 7.2.

Example 7.3 (continued): Using the Matrix D of Example 7.3, algorithm (7.1) was
simulated for various values of θ. For each θ, 50 sample runs were made and
the number of times the actual sample paths converge to the neighborhood of different
solution points are given in the following Table 7.1

Table 7.1

a = 0.1			b = 0.001

Solutions of $W_i(p) = 0$ $i = 1,2$	The process $\{P(k)\}$ converge to different solutions of $W_i(P) = 0, i - 1,2$		
	$\theta = 1.0$	$\theta = 0.5$	$\theta = 0.25$
$p^a \rightleftharpoons$ (0,0)	43	48	50
$p^c \simeq$ (1,1)	7	2	0

The actual sample paths in two typical cases are plotted in Figure 7.7 for
$\theta = 0.5$. The above simulation results lead to the conjecture that the conclusions
of Theorem 1 must be true despite the fact that there are multiple solutions.
In fact, the mathematical characterization of $\eta(k)$ when $W_i(P) = 0, i = 1,2$ have
multiple solutions remains an open problem.

Remark 7.7: One special case deserves mention here. In addition to (7.11), if
$d_{11} = d_{12}$ and $d_{21} = d_{12}$ it can be seen that $W_i(P)$, $i = 1,2$ are symmetric in p_1
and q_1 and that $W_i(P^*) = 0$ for $P^* = (1/2, 1/2)$ for all values of a and b. And
there exists b^* such that for all $0 < b < b^*$, $W_i(P) = 0$ exhibits two more solutions:
one near (0,0) and another near (1.1). (See Exercise 7.7)

7.5   COMMENTS AND HISTORICAL REMARKS:

The concept of Team was introduced by Marschak in 1955 and the book by Marschak and Radner [M2] gives a detailed treatment of various topics on multipers team decision theory.   Since this pioneering work, a number of papers have appeared in systesm literature on team decision making, decentralized control, structure and transfer of information between various members of the team, etc. [H5] [H6] to mention a few.

During the 60's and early 70's, there appeared a variety of papers dealing with multimodal optimization using stochastic automation models. [C4], [f4], [M3], [M4], [V5], [V4]. Using the language of this chapter they can all be paraphrased as decentralized team decision problems.

Most of the contents of this chapter follows [L10]. We hope that this chapter will motivate further research on the learning approach to decentralized team decision making.

7.6 EXERCISES:

7.1 Show that all the states of the Markov process $\{P(k)\}$ under the algorithm
(7.1) are non-absorbing when $0 < d_{ij} < 1$ and $0 < \theta_i^A$, $\theta_j^B < 1$, $i,j = 1,2$.

7.2 Compute the eigenvalues of $W'(P*)$ exactly and verify the accuracy of the
approximations used in this chapter.

7.3 Instead of the L_{R-P}^E algorithm (7.1), start with the N_{R-P}^E algorithm (refer to
algorithm 2.1) and verify the conclusions of Section 7.2.

7.4 Consider the following 3 x 3 matrix:

$$D = \begin{bmatrix} .6 & .3 & .9 \\ .1 & .5 & .4 \\ .8 & .2 & .7 \end{bmatrix}$$

It can be seen that this is a multimodal matrix with modes at (1,1), (1,3),
(2,2), (3,1), and (3,3).

a) Give a (reasonable) definition of a mode for a general matrix of order
M x N (M, N $\geq$ 2).

b) Give conditions for a general M x N matrix to be unimodal.

c) How many modes are possible in a general matrix of order M x N?

7.5 Following Theorem (5.6) prove all the conclusions (7C.1) - (7C.3).

7.6 Following the suggestions of Remark (7.6) show that (7.10) holds when $d_{21} <$
$d_{22} < d_{12}$.

7.7 Verify the conclusions in Remark 7.7

7.8 Extend the results of this chapter to general non-zero sum game problems
with incomplete information.

CHAPTER 8

Control of a Markov Chain with Unknown

Dynamics and Cost Structure

8.1 Introduction:

This chapter deals with the application of the "absolutely expedient"
learning algorithms (developed in chapter 3) for the problem of control of
a finite state Markov chain whose transition probabilities as a function
of a finite number of control actions are unknown. At any instant of time
depending on the state of the Markov chain and the control action chosen a
reward is incurred. It is assumed that this reward is a two valued (binary)
random variable whose distribution as a function of the state and the control action is unknown, but the sequence of states actually visited by the
Markov chain is available. In other words we consider a Markov chain whose
dynamics and reward structure are unknown but the state is observable
exactly.

We associate with each state a learning algorithm which updates the
distribution from which the control actions are chosen. It is shown that
if the underlying structure satisfies the usual conditions [V1] for the
existence of single ergodic class there exists a proper choice of para-

meters of the learning algorithm such that we will eventually come up with the one step optimal control vector with probability as close to unity as desired.

Basic definitions and precise statement of the problem are given in section 8.2. The learning algorithm and the analysis are presented in section 8.3 and 8.4 respectively. While section 8.5 presents a variety of simulation, in section 8.6 we consider an extension of these ideas to the case when the state information of the Markov chain is available exactly but with unit delay. In this case to make the problem tractable we assume the dynamics are known but cost structure unknown. A new class of learning algorithms is discussed in this section.

8.2 Definitions and Statement of Problem:

Consider a finite Markov chain with $X = \{x_1, x_2, \ldots, x_K\}$, $(2 \leq K < \infty)$ as its state space. When in each state, one of $M(2 \leq M < \infty)$ possible control actions can be chosen from the set $U = \{\alpha_1, \alpha_2, \ldots, \alpha_M\}$.

a) Reward Structure: Let H: $X \times U \rightarrow R$, the set of all real numbers where

$$H[x_i, \alpha_r] = +1 \text{ (unit gain) with probability } d_{ir}$$
$$= -1 \text{ (unit loss) with probability } c_{ir}$$

and $c_{ir} + d_{ir} = 1$ for all i = 1,2,...,K, r = 1,2,...,M   (8.1)

If we denote the state at time k as x(k) and the control action chosen at time k as $\hat{u}(k)$, with a slight abuse of notation let

$$H(k) \triangleq H[x(k), \hat{u}(k)] \tag{8.2}$$

The expected reward (or gain) at time k is given by

$$E[H(k) \mid x(k) = x_i, \hat{u}(k) = \alpha_r] = 2d_{ir} - 1 \tag{8.3}$$

Assumption 8A.1: It is assumed that d_{ir} are <u>unknown</u> for all i and r. Further $0 < d_{ir} < 1$ for all i and r and d_{ir} are distinct for each i.

b) <u>Control policy</u>: At time k, suppose that the control action is chosen randomly from a distribution defined below:

$$\left. \begin{array}{l} \text{Prob } [\hat{u}(k) = \alpha_r \mid x(k) = x_i] = p_r^{(i)}(k) \\[2mm] \text{where} \\[1mm] \sum_{r=1}^{M} p_r^{(i)}(k) = 1, \; 0 \leq p_r^{(i)}(k) \text{ for all } k \\[2mm] p^{(i)}(k) = (p_1^{(i)}(k), \; p_2^{(i)}(k) \; \ldots \ldots \; p_M^{(i)}(k) \;)^T \end{array} \right\} \tag{8.4}$$

Thus $p^{(i)}(k) \in S_M$, the M-dimensional unit simplex for all k. Let $V_M = \{e_1, e_2, \ldots, e_M\}$ be the set of all unit vectors forming the vertices of the simplex S_M.

A <u>control policy</u> R is a sequence of a set of probability vectors $\{p^{(i)}(k): i = 1,2,\ldots,K\}$, $k \geq 0$ where $p^{(i)}(k)$ is defined as in (8.4). Such a control policy is in general called a "<u>Nonstationary Random Markov Policy</u>" (NSRMP)[1] as the probability of choosing an action depends on x(k) as well as on k. If $p^{(i)}(k) \equiv p^{(i)}$ for all i = 1,2,\ldots,K, then $\{p^{(k)}: i = 1,2,\ldots,K\}$ is called "<u>Stationary Random Markov Policy</u>," (SRMP). In addition, if each $p^{(i)}$ is a unit vector (one component unity, rest are all zeros) then the corresponding policy is called "<u>Stationary Deterministic Markov Policy</u>" (SDMP). In the latter case, we associate a control action u^i with state x_i.

---

[1] By NSRMP we will also mean, in the following, the set of all non stationary random Markov policies. Similarly for other class of policies.

Any policy $R \in SDMP$ is uniquely determined by a control vector

$$\underline{u} = (u^1, u^2, \ldots, u^K)^T, \quad u^i \in U \text{ for all } i = 1, 2, \ldots, K \text{ and is denoted}$$

as $R = \{\underline{u}\}$. Let $L = \{\underline{u} \mid \underline{u} = (u^1, u^2, \ldots, u^K)^T, \quad u^i \in U, \quad i = 1, 2, \ldots, K\}$

denote the set of all SDMPs. Let $\underline{u}(k) = (u^1(k), u^2(k) \ldots u^K(k))^T$ and

$\underline{\hat{u}}(k) = (\hat{u}^1(k), \hat{u}^2(k), \ldots, \hat{u}^K(k))^T$ denote the control vectors corres-

ponding to non-stationary DMPs and NSRMPs respectively where $u^i(k)$

and $\hat{u}^i(k)$ refer to the control action chosen when $x(k) = x_i$ under

these policies.$^{(2)}$

c) Transition Structure:

Define the matrix

$$P[\underline{u}] = [p_{ij}[u^i]] \tag{8.5}$$

where

$$p_{ij}[u^i] = \text{Prob} \ [x(k+1) = x_j \mid x(k) = x_i, \ u^i]$$

$$0 < p_{ij} \ [u^i], \sum_{j=1}^{K} p_{ij} \ [u^i] = 1, \ i = 1, 2, \ldots, K$$

Let $P = \{P[u] \mid \underline{u} \in L\}$ be the set of all stochastic matrices corres-

ponding to all possible SDMPS.

The transition matrix $P[\underline{\hat{u}}(k)] = p_{ij}[\hat{u}^i(k)]$ under a NSRMP is

given by

$$p_{ij}[\hat{u}^i(k)] = \sum_{r=1}^{M} p_r^i(k) \ p_{ij} \ [\hat{u}^i(k) = \alpha_r] \tag{8.6}$$

Evidently, $P[\underline{\hat{u}}(k)]$ belongs to the class of matrices whose columns are

in the convex hull generated by the columns of the elements in P.

Assumption 8A.2: The set P is unknown but it is assumed that

$$p_{ij} \ [u^i] \geq \delta \ > \ 0 \tag{8.7}$$

for all $i, j = 1, 2, \ldots, K$ and $u^i \in U$

---

(2) Notice $\hat{u}(k)$ just refers to the control action chosen at time k but $\hat{u}^i(k)$
refers to the state and the control action at time k.

Define (with the slight abuse of notation)

$$\Pi_i(k) = \text{Prob } [x(k) = x_i]$$

and

$$\Pi(k) = (\Pi_1(k), \Pi_2(k), \ldots, \Pi_K(k))^T \qquad (8.11)$$

where

$$\Pi^T(k+1) = \Pi^T(k) P[\hat{u}(k)]$$

with $\Pi(0) \in S_K$, the K-dimensional unit simplex. (See Exercise 8.2)

d) Performance Index:

The long run expected reward per unit time for any policy R is given by

$$J_R = \lim_{N \to \infty} \frac{1}{N+1} E_R \{ \sum_{k=0}^{N} H[x(k), \hat{u}(k)] \} \qquad (8.12)$$

where E_R is the expectation taken with respect to all possible choices of states, actions and random reward (unit gain or loss) under the policy R.

If $R \in$ SDMP, that is, $R = \{\underline{u}\}$ then (8.12) becomes (from Exercise (8.2) and (8.1))

$$J_R \triangleq J(\underline{u}) = \lim_{N \to \infty} \frac{1}{N+1} \sum_{k=0}^{N} \Pi^T(0) P^k[\underline{u}] E[\underline{H}[\underline{u}]]$$

$$= \Pi^T(\underline{u}) E[\underline{H}[\underline{u}]] \qquad (8.13)$$

where $\Pi(u)$ satisfies (8.9) and

$$\underline{H}[\underline{u}] \triangleq (H(x_1, u^1), H(x_2, u^2) \ldots H(x_K u^K))^T \qquad (8.14)$$

This assumption implies [K8] that $P[\underline{u}]$ for each $\underline{u} \in L$ is regular and the system

$$\Pi^T(\underline{u}) = \Pi^T(\underline{u}) P[\underline{u}] \qquad (8.8)$$

has a unique solution $\Pi(\underline{u})$ where

$$\Pi(\underline{u}) = (\Pi_1(\underline{u}) \Pi_2(\underline{u}), \ldots, \Pi_K(\underline{u}))^T$$

$$0 < \Pi_i(\underline{u}), \sum_{i=1}^{K} \Pi_i(\underline{u}) = 1. \qquad (8.9)$$

Further it is known (Exercise 8.1) that

$$\lim_{N \to \infty} P^N[\underline{u}] = \underline{e} \; \Pi^T(\underline{u}) \tag{8.10}$$

where $\underline{e} = (1,1,\ldots,1)^T$ [3]

By assumption 8A.1, d_{ir} are distinct for each i and since $E[H[\mathbf{X}_i, \alpha_r]] = 2d_{ir} - 1$, there exists a unique i_* for each i such that

$$d_{ii_*} = \max_r \{d_{ir}\} \tag{8.15}$$

This naturally leads to the definition of <u>one step optimal control vector</u> $\underline{u}_* = (u^1_*, u^2_* \ldots u^K_*)$ where $u^i_* = \alpha_{i_*}$ for all i = 1, 2, ... **K**. Further, since there are only finitely many SDMPs, there exists an unique $\underline{u}_{opt} = (u^1_{opt}, u^2_{opt} \ldots u^K_{opt})$ called <u>optimal control</u> vector where

$$\left. \begin{array}{l} J(\underline{u}_{opt}) \geq J(\underline{u}) \quad \text{for all } \underline{u} \in \mathbf{L} \\ \\ \\ J(\underline{u}_{opt}) \geq J(\underline{u}_*) \end{array} \right\} \tag{8.16}$$

and

The following example illustrates the above concepts.

<u>Example 8.1</u>:- Let X = $\{\mathbf{X}_1, \mathbf{X}_2\}$ and $\mathbf{U} = \{\alpha_1, \alpha_2\}$

Reward Structure

State	Probabilities d_{ir}	
	Action α_1	Action α_2
$\mathbf{X}_1$	0.9	0.5
$\mathbf{X}_2$	0.4	0.7

[3] P^N refers to the Nth power of the matrix $P[\underline{u}]$.

Dynamics

From State	Action	Transition Probabilities	
		To State X_1	To State X_2
X_1	α_1	0.9	0.1
	α_2	0.1	0.9
X_2	α_1	0.8	0.2
	α_2	0.2	0.8

The following computations are easily verified

$\underline{u}$	$J(\underline{u})$
$\underline{u}_{opt} = (\alpha_1, \alpha_1)$	0.844
$\underline{u}_* = (\alpha_1, \alpha_2)$	0.833
$\underline{u} = (\alpha_2, \alpha_2)$	0.745
$\underline{u} = (\alpha_2, \alpha_1)$	0.447

Two new classes of policies are now introduced.

Definition 8.1:- A policy $R = \{p^{(i)}(k): i = 1, 2, \ldots k\} \in$ NSRMP
is said to be an **asymptotically stationary random Markov policy** (ASRMP) if
for each i ($= 1, 2, \ldots k$)

$$\text{Prob}\left[\lim_{k \to \infty} p^{(i)}(k) = e_j\right] = r_j^{(i)} \qquad (8.17)$$

where

$$0 < r_j^{(i)} \text{ and } \sum_{j=1}^{M} r_j^{(i)} = 1 \text{ and } e_j \text{ is the } j^{th} \text{ unit vector.}$$

Definition 8.2:- A policy $R \in$ ASRMP is said to be one step ε - optimal
(ε - optimal) for $\varepsilon > 0$ if

$$J_R \geqslant J(\underline{u}_*) - \varepsilon \ (J_R \geqslant J(\underline{u}_{opt}) - \varepsilon) \qquad (8.18)$$

where $\underline{u}_*$ ($\underline{u}_{opt}$) is defined above.

Assumption 8A.3:- The state of the Markov chain is known exactly at each
instant of time that is, x(k) is known for each $k \geq 0$

e) Statement of Problem:- under the assumptions 8A.1 - 8A.3 our aim in
this chapter is to establish on-line learning algorithm which yields one
step ε - optimal policies.

Remark 8.1:- Notice that the one step optimal control vector $\underline{u}_*$ depends
only on the reward structure but the optimal control vector $\underline{u}_{opt}$ depends
on both the dynamics and reward structure. At present time we are only
able to find the one step ε - optimal control vector. This is primarily
because the learning algorithms of interest are insensitive to transition

probabilities. The design of learning algorithms which are sensitive to reward structure and dynamics in such a way as to find ε - optimal (instead of one step ε - optimal) control vector constitutes an important open problem in this area.

8.3 Learning Algorithm:

Specifying a NSRMP is equivalent to specifying a sequence of a set of probability distributions $\{p^{(i)}(k): i = 1,2,\ldots,K\}$, $k \geq 0$ where $p^{(i)}(k)$ is defined in (8.4). At time k if $x(k) = x_i$, the action $\hat{u}^i(k) = \alpha_j$ is chosen from the distribution $p^{(i)}(k)$. The effect of α_j on x_i is to cause a unit gain with probability d_{ij} and unit loss with probability $c_{ir} = 1 - d_{ir}$. If the values of d_{ij}'s are known, then the sequence of probability vectors maximizing J_R can be calculated using the well known techniques [B3] [V1] [D1] [K9][H8]. In the absence of such information, we propose to update $p^{(i)}(k)$ as follows: If $x(k) = x_i$, then

$$\left.\begin{array}{l} p^{(m)}(k+1) = p^{(m)}(k) \text{ for all } m \neq i \\ p^{(i)}(k+1) = T^{(i)}[p^{(i)}(k),\ \hat{u}(k),\ H(k)] \end{array}\right\} \tag{8.19}$$

where

$$T^i: S_M \times \mathcal{U} \times \{-1, +1\} \to S_M$$

The specific details of T^i are given below:

If $x(k) = x_i$, $\hat{u}(k) = \alpha_j$

$$p^{(i)}_j(k+1) = \begin{cases} p^{(i)}_j(k) + \lambda^{(i)}[p^{(i)}(k)] \; [1-p^{(i)}_j(k)] & \text{if } H(k) = +1 \\ p_j(k) - \mu^{(i)}[p^{(i)}(k)] \; [1 - p^{(i)}_j(k)] & \text{if } H(k) = -1 \end{cases}$$

(8.20)

$$\begin{array}{c} p^{(i)}(k+1) \\ (\ell \neq j) \end{array} = \begin{cases} p^{(i)}_\ell(k) - \lambda^{(i)}[p^{(i)}(k)] \; p^{(i)}_\ell(k) & \text{if } H(k) = +1 \\ p^{(i)}_\ell(k) + \mu^{(i)}[p^{(i)}(k)] \; p^{(i)}_\ell(k) & \text{if } H(k) = -1 \end{cases}$$

where $\lambda^{(i)}: S_M \to [0,1]$, $\mu^i: S_M \to [0,1]$ are such that

(8C.1) $0 \leq \lambda^{(i)}[p] < 1$ with $\lambda^{(i)}[p] = 0$ only if $p \in V_M$

(8C.2) $\mu^{(i)}[p] \equiv 0$ or $0 \leq \mu^{(i)}[p] < \min\limits_{r} \left\{ \dfrac{p_r}{1-p_r} \right\}$

for $0 < p_r < 1$, $r = 1, 2, \ldots, M$

Remark 8.2: The above algorithm (8.20) is called the <u>absolutely expedient</u> algorithm and has been extensively studied in chapter 3. There are totally K algorithms, each being associated with one of the K states of the Markov chain. At any time as the system can be in only one of its states, only one of these K algorithms is active. Conditions (8C.1) – (8C.2) insure that $p^{(i)}(k+1) \in S_M$ if $p^{(i)}(k)$ does. Further, only for simplicity in presentation did we assume that the number of control actions is the same for all states. Virtually every conclusion will hold even when we allow the number of permissible control actions to be different for different states.

8.4 Analysis:

As a first step let us investigate the implications of the assumption 8A.2. To this end define

$$B(k) \triangleq \prod_{m=0}^{k} P[\hat{\underline{u}}(m)] \tag{8.21}$$

$$= P[\hat{\underline{u}}(k)] \ P \ [\hat{\underline{u}}(k-1)] \ \cdots \cdots \ P[\hat{\underline{u}}(0)]$$

that is, $B(k)$ is the backward product of transition matrices. Let $B(k)=[b_{ij}(k)]$. We immediately have the following

Lemma 8.1: Under the assumption 8A.2, for all $k \geq 1$

$$\Pi_j(k) = \text{Prob}[x(k) = x_j] \geq \delta > 0 \tag{8.22}$$

Proof: From (8.11)

$$\Pi^T(k+1) = \Pi^T(k) \ P[\hat{\underline{u}}(k)]$$
$$= \Pi^T(0) \ B^T(k)$$

The lemma follows from exercises (8.4)   Q.E.D.

Stated in words, the process $x(k)$ visits each state with a non zero probability for all $k \geq 1$. Define $y(k) = (p^{(1)}(k), p^{(2)}(k), \ldots, p^{(K)}(k))$. Clearly $\{x(k), y(k)\}$ $k \geq 0$ is a stationary Markov process with $X \times S_M^K$ as its state space where S_M^K refers to the K-fold Cartesian product of the unit simplex S_M.

Define

$$h_i(k) = \sum_{r=1}^{M} p_r^{(i)}(k) \ d_{ir} \tag{8.23}$$

Our first result is concerning the behavior of the sequence $\{h_i(k)\}$

Theorem 8.1: If the set of probability vectors $\{p^{(i)}(k): i=1,2,\ldots,K\}$ is

updated according to (8.19) - (8.20) and if $p^{(i)}(0) \notin V_M$ for all

$i=1,2,\ldots,K$, then

$$E[h_i(k+1) \mid h_i(\ell) : \ell = 1,2,\ldots,k] \geq h_i(k) \tag{8.24}$$

with probability one for all $i = 1,2,\ldots,K$. Stated in other words for

each i, $\{h_i(k)\}$ is a non-negative bounded submartingale.

Proof: Consider

$$E[h_i(k+1) - h_i(k) \mid y(\ell) : \ell = 1,2,\ldots,k]$$

$$= \sum_{j=1}^{M} \text{Prob} [x(k) = x_j \mid y(\ell): \ell = 1,2,\ldots,k] \, \delta h_i^j(k) \tag{8.25}$$

where

$$\delta h_i^j(k) \triangleq E[h_i(k+1) - h_i(k) \mid x(k) = x_j, y(\ell) : \ell = 1,2,\ldots,k] \tag{8.26}$$

But from the algorithm (8.19) - (8.20) it follows that

$$\delta h_i^j(k) = 0 \quad \text{for } i \neq j \tag{8.27}$$

Thus

$$E [h_i(k+1) - h_i(k) \mid y(\ell) : \ell = 1,2,\ldots,k]$$
$$= \text{Prob} [x(k) = x_j \mid y(\ell): \ell = 1,2,\ldots,k] \, \delta h_i^i(k) \tag{8.28}$$

The right hand side of (8.28) is the product of two terms of which the

first term, by lemma 8.1, is bounded away from zero. It can be shown

that (Exercise 8.5)

$$\delta h_i^i(k) = \frac{1}{2} \{\lambda^{(i)} [p^{(i)}(k)] + \mu^{(i)} [p^{(i)}(k)]\} p^{(i)^T}(k) \, D^{(i)} \, p^{(i)}(k) \tag{8.29}$$

where the matrix $D^{(i)}$ is given by

$$D^{(i)} = [D_{rm}^i] = [(d_{im} - d_{ir})^2] \tag{8.30}$$

From conditions (8C.1) - (8C.2) and distinctness of d_{ir} stated in assumption (8A.1) we obtain that

$$\delta h_i^i (k) \geq 0 \qquad (8.31)$$

with equality holding only when $p^{(i)}(k) \in V_M$. Now taking conditional expectations of both sides of (8.28), by the law of iterated conditional expectations and from the above observations, we easily obtain (8.24)

Q.E.D.

As a consequence we have the following:

Theorem 8.2: $\lim_{k \to \infty} p^{(i)}(k)$ exists and is an element of V_M with probability one for all $i = 1, 2, \ldots, K$.

Proof: Since $\{h_i(k)\}$ is a submartingale, by martingale theorems [D 2] [T7] it follows that $\lim_{k \to \infty} h_i(k)$ and hence $\lim_{k \to \infty} p^{(i)}(k)$ exists with probability one. Also $\lim_{k \to \infty} E[h_i(k+1) - h_i(k) \mid h_i(\ell): \ell = 1, 2, \ldots, k] = 0$ with probability one. From (8.29), the latter is true only if either

a) $\lim_{k \to \infty} [\lambda^{(i)}[p^{(i)}(k)] + \mu^{(i)}[p^{(i)}(k)]] = 0$

or

b) $\lim_{k \to \infty} [p^{(i)^T}(k) \; D^{(i)} \; p^{(i)}(k)] = 0$

$$(8.32)$$

with probability one. Again from (8C.1) - (8C.2) and distinctness of d_{ir} (for each i) we see that either of the above events can take place only if $\lim_{k \to \infty} p^{(i)}(k) \in V_M$ with probability one. Q.E.D.

Corollary 8.1: Every NSRMP, R generated by the algorithm (8.19) - (8.20) is in fact an ASRMP.

Proof: It follows from theorem 8.2 that $\lim_{k \to \infty} y(k)$ exist and is an element of V_M^K (K-fold Cartesian product of V_M) with probability one. Hence the corollary by the definition of ASRMP.   Q.E.D.

Recall that the algorithm (8.19) - (8.20) updates $p^{(i)}(k)$ only at instances when $x(k) = x_i$. From lemma 8.1, we know that the latter event occurs with a non zero probability. Let k_0^i, k_1^i, k_2^i ... be the actual instances at which $x(k) = x_i$. It is well known [8] that $\{k_t^i\}$, $t \geq 0$ is a renewal sequence, that is $\{(k_{t+1}^i - k_t^i)\}$, $t \geq 1$ is a sequence of independent random variables.

Define

$$\tilde{p}^{(i)}(t) = p^{(i)}(k_t^i) \quad \text{for } t = 0,1,2,\ldots \tag{8.33}$$

These observations lead to the following conclusions:

   a) $p^{(i)}(k)$ is a semi-Markov process and $\tilde{p}^{(i)}(t)$ is an imbedded

       (stationary) Markov process

   b) The Markov process $\tilde{p}^{(i)}(t)$ is such that $\lim_{t \to \infty} \tilde{p}^{(i)}(t)$ exists and

       belongs to V_M with probability one (Exercise (8.6)

Thus $\{\tilde{p}^{(i)}(t)\} t \geq 0$ belongs to the class of processes generated by "absolutely expedient" learning algorithms whose asymptotic behavior has been extensively studied in chapter 3. To save space we only indicate the major steps in the further analysis of the process $\{\tilde{p}^{(i)}(t)\}$ and hence of $\{\tilde{p}^{(i)}(k)\}$. For details refer chapter 3. Let

$$\Gamma_j^{(i)}(p) \triangleq \text{Prob} \left[\lim_{t \to \infty} \tilde{p}^{(i)}(t) = e_j \mid \tilde{p}^{(i)}(0) = p \right] \tag{8.34}$$

Then, from the above conclusions it is clear that

$$\sum_{j=1}^{M} \Gamma_j^{(i)}(p) = 1 \quad \text{for all } i = 1,2,\ldots,K \tag{8.35}$$

Combining (8.16) and (8.34) we see that $\Gamma_{i_*}^{(i)}(p)$ is the probability that asymptotically the (one step ε-optimal) control action α_{i_*} will be chosen whenever the Markov chain is in state x_i. The following theorem gives a lower bound on this probability.

__Theorem 8.3:__ For each $\delta_i > 0$ there exists a $z_i > 0$ and a choice of functions $\lambda^{(i)}[.]$ and $\mu^{(i)}[.]$ in the algorithm (8.20) such that

$$1 - \delta_i \leq \frac{1-e^{-p_{i_*}z_i}}{1-e^{-z_i}} \leq \Gamma_{i_*}^{(i)}(p) < 1 \tag{8.36}$$

__Remark 8.3:__ The proof of this theorem is very lengthy and follows very closely the developments in chapter 3 especially those in section 3.4. We invite the reader to fill in the necessary details. The choice of functions referred to in the statement of the theorem is obtained by replacing $\lambda^{(i)}[.]$ and $\mu^{(i)}[.]$ by $\theta\lambda^{(i)}[.]$ and $\theta\mu^{(i)}[.]$ where $0 < \theta < 1$. θ is known as the step length parameter (refer section 3.6) Norman [N11] originally derived the bounds on $\Gamma_{i_*}^{(i)}(p)$ for the special case when $\lambda^{(i)}(p) \equiv a$, $0 < a < 1$ and $\mu^i(p) \equiv 0$. The above generalization is due to Lakshmivarahan and Thathachar [L 5].

An immediate consequence of the above theorem is the following:

__Corollary 8.3:__ For each $\varepsilon_i > 0$ there exists a choice of $\lambda^{(i)}[.]$ and $\mu^{(i)}[.]$ in the algorithm (8.20) such that

$$\lim_{k\to\infty} E[h_i(k)] > d_{ii_*} - \varepsilon \tag{8.37}$$

__Proof:__

$$\lim_{k\to\infty} E[h_i(k) \mid p^{(i)}(0) = p] = \sum_{r=1}^{M} \lim_{k\to\infty} E[p_r^{(i)}(k) \mid p^{(i)}(0) = p]d_{ir}$$

$$= \sum_{r=1}^{M} \Gamma_r^{(i)}(p) \, d_{ir}$$

$$\geq d_{ii_*} - \delta_i \sum_{r\neq i_*} (d_{ii_*} - d_{ir})$$

$$> d_{ii_*} - \varepsilon_i \tag{8.38}$$

where the second equality follows from the fact $\tilde{p}^{(i)}(t)$ and $p^{(i)}(k)$ have the same asymptotic properties and the last inequality follows by choosing

$$\delta_i < \frac{\varepsilon_i}{\sum\limits_{r \neq i_*} (d_{ii_*} - d_{ir})} \tag{8.39}$$

Notice $\sum\limits_{r \neq i_*} (d_{ii_*} - d_{ir}) > 0$. Now taking expectations of both sides of (8.38), the corollary follows.   Q.E.D.

The above theorem 8.3 enables us to strengthen the conclusions of the corollary 8.1 as in the following theorem.

Theorem 8.4: For every $\delta > 0$, there exists proper choice of functions $\lambda^{(i)}$ [.] and $\mu^{(i)}$ [.] in (8.20) such that the sequence of control vectors $\{\hat{u}(k)\}$ generated by the algorithm (8.19) - (8.20) converges to the one step ε-optimal control vector $\underline{u}_*$ with probability greater than or equal to $(1 - \delta)$.

Proof: Recall that $\hat{\underline{u}}(k) = (\hat{u}^1(k), \hat{u}^2(k) \ldots, \hat{u}^K(k))^T$ where $\hat{u}^i(k)$ is randomly chosen from the distribution $p^{(i)}(k)$. It follows from theorem 8.3 that $\hat{u}^i(k)$ converges to $u_*^i = \alpha_{i_*}$, the optimal action for the state x_i, with probability greater than or equal to $(1-\delta_i)$. Now choose $\delta_i > 0$ in (8.36) such that $(1-\delta) \geq \prod\limits_{i=1}^{M}(1-\delta_i)$ to obtain the desired result. Q.E.D.

We now state the main result of this chapter.

Theorem 8.5: For every $\varepsilon > 0$ there exists proper choice of parameters of the algorithm (8.19) - (8.20) such that every policy $R \in$ ASRMP generated by the algorithm is one step ε-optimal, that is, $J_R \geq J(\underline{u}_*) - \varepsilon$

Proof: From theorem 8.4 it follows that the sequence of transition matrices $P[\hat{\underline{u}}(k)]$ is such that $\lim\limits_{k \to \infty} P[\hat{\underline{u}}(k)] = P[\underline{u}_*]$ with probability $)1 - \delta)$.

Similarly, with the same probability $\lim\limits_{k \to \infty} \Pi(k) = \Pi[\underline{u}_*]$. These in turn imply that with probability greater than or equal to $(1 - \delta)$

$$\lim\limits_{k \to \infty} E_R [H(x(k), \hat{u}(k)] = \Pi^T[\underline{u}_*] E [\underline{H}[\underline{u}_*]] \tag{8.40}$$

where $\underline{H}$ [.] is defined in (8.14).

Hence

$$J_R = \lim_{N \to \infty} \frac{1}{n+1} \ E_R \sum_{k=0}^{N} H[x(k), \hat{u}(k)]$$

(4)

also converges to the same limit $\Pi^T[\underline{u}_*] \ E \ [\underline{H} \ [\underline{u}_*]]$ with probability $\geq (1-\delta)$.

Thus

$$J_R \geq (1-\delta) \ \Pi^T[\underline{u}_*] E[\underline{H}[\underline{u}_*]] + \delta \max_{u \neq u_*} \{\Pi^T[\underline{u}] \ E \ [\underline{H} \ [u]]\}$$

$$= \ \Pi^T[\underline{u}_*] \ E[\underline{H}[\underline{u}_*]] \ -\delta[\Pi^T[\underline{u}_*]E \ [\underline{H}[\underline{u}_*]] \ - \max_{u \neq u_*} \left\{\Pi^T[\underline{u}]E[\underline{H}[\underline{u}]]\right\} \ ]$$

$$\geq J(\underline{u}_*) - \varepsilon$$

where the coefficient of δ in the last but one inequality is positive and

the last inequality follows by proper choice of δ.   Q.E.D.

8.5 Simulations:

In order to provide some idea of the nature of convergence that can

be obtained with the algorithms proposed a 3-state Markov chain was simu-

lated. The number of actions was assumed to be 2 for state x_1 and 3 for

states x_2 and x_3. The other details are given below in Tables 8.1 and

8.2.

Table 8.1

States	Probabilities (d_{ir})		
	Action α_1	Action α_2	Action α_3
x_1	.6	.3	—
x_2	.4	.7	.3
x_3	.1	.45	.8

(4)Here we use the fact if a sequence converges then the Cesaro sum converges
to the same limit.

Table 8.2

From State	Action	Transition Probabilities		
		To State x_1	To State x_2	To State x_3
x_1	α_1	0.6	0.3	0.1
	α_2	0.5	0.1	0.4
x_2	α_1	0.2	0.6	0.2
	α_2	0.1	0.8	0.1
	α_3	0.2	0.5	0.3
x_3	α_1	0.1	0.2	0.7
	α_2	0.2	0.1	0.7
	α_3	0.4	0.1	0.5

From an examination of Table 8.1 one can read off the one step optimal action associated with each state and also the minimum expected cost which results from such a choice. Table 8.3 displays this information.

Table 8.3

State	Optimal Control Action	Maximum Expected Reward
x_1	α_1	.6
x_2	α_2	.7
x_3	α_3	.8

It is of course assumed that the data in Table 8.1 is not available and Table 8.3 just serves as a standard for comparison of the behavior of the learning algorithm. Two learning algorithms of the type mentioned in section 8.3 were used.

Algorithm I: $\lambda^{(i)}(\underline{p}) = 0.5$ for all $i = 1,2,,3$

$$\mu^{(i)}(\underline{p}) = 0.1\ p_1^{(i)} p_2^{(i)} \cdots p_M^{(i)}$$

where M = 2   for $i = 1$

      = 3   for $i = 2,3$

Algorithm II: $\lambda^{(i)}(\underline{p}) = 0.5$

$$\mu^{(i)}(\underline{p}) = 0$$

Using each of the above algorithms fifty sample runs were made and the reward obtained in each state over the fifty runs was computed at several instants. The average cost is represented by $E_R[H[x_i, \hat{u}(k)]]$ are given in Tables 8.4 and 8.5.

It is seen from these tables that the expected reward in each state converges very close to the maximum value possible. It follows that J_R defined in (8.12) converges very close to its maximum value. Algorithm I is faster than Algorithm II as is to be expected, since the penalty term $\mu^{(i)}$ is zero in the latter.

Table 8.5

Algorithm II

k	$E_R[H[x_1,\hat{u}(k)]]$	$E_R[H[x_2,\hat{u}(k)]]$	$E_R[H[x_3,\hat{u}(k)]]$
0	0.4500	0.4333	0.4500
100	0.4877	0.5037	0.5911
200	0.5172	0.5700	0.6626
300	0.5426	0.6259	0.7111
400	0.5625	0.6590	0.7359
500	0.5747	0.6753	0.7548
600	0.5825	0.6875	0.7690
700	0.5862	0.6936	0.7801
800	0.5881	0.6975	0.7838
900	0.5939	0.6992	0.7887
1000	0.5949	0.6999	0.7921
1100	0.5967	0.7000	0.7944
1200	0.5980	0.7000	0.7967
1300	0.5988	0.7000	0.7971
1400	0.5993	0.7000	0.7979
1500	0.5997	0.7000	0.7989

Table 8.4

Algorithm I

k	$E_R[H[x_1,\hat{u}(k)]]$	$E_R[H[x_2,\hat{u}(k)]]$	$E_R[H[x_3,\hat{u}(k)]]$
0	0.4500	0.4333	0.4500
100	0.4993	0.5110	0.5981
200	0.5371	0.5785	0.6728
300	0.5604	0.6336	0.7171
400	0.5759	0.6632	0.7379
500	0.5866	0.6794	0.7577
600	0.5921	0.6900	0.7740
700	0.5951	0.6944	0.7829
800	0.5948	0.6979	0.7879
900	0.5973	0.6986	0.7916
1000	0.5985	0.6999	0.7952
1100	0.5989	0.7000	0.7969
1200	0.5991	0.7000	0.7981
1300	0.5993	0.7000	0.7981
1400	0.5994	0.7000	0.7985
1500	0.5998	0.7000	0.7993

8.6 Extension to delayed state observation:

In this section we consider the problem of control of a Markov chain when the state process x(k) is known exactly but with one step delay. For later reference let us begin by listing all the assumptions to be used in this section.

(8A'.1) d_{ir} are unknown, $0 < d_{ir} < 1$ for all i and r. Also d_{ir} are distinct for each i. (Notice this is same as 8A.1 and is given here only for completeness).

(8A'.2) The set P is known and

$$p_{ij}[u^i] \geq \delta > 0$$

for all $i,j = 1,2,\ldots,K$, $u^i \in \mathcal{U}$

(8A'.3) The state x(k) is known exactly but with one step delay. That is, if z(k) is the observed process then z(k) = x(k-1) with probability one.

Our aim is to find, under assumptions (8A'.1) - (8A'.3), an online an learning lgorithm which yields one step ε-optimal policies. For this case of delayed state observation even with the assumption 8A'.2 that p[u] is known for all u ε L, it will be seen below that the analysis gets very complex and involved. The problem remains to be solved when in addition P is also unknown.

As a motivation for the algorithm to be developed here, we first summarize the sequence of steps in the learning process when the state observation is available without delay described in the sections 8.3 and 8.4.

Step 1: The system is observed at time k. Let x(k) = x_i.

Step 2: A control action is picked from the set $\mathcal{U}$ on the basis of the distribution $p^{(i)}(k)$ where $p^{(i)}(k)$ is defined in (8.4).

<u>Step 3</u>: $p^{(i)}(k)$ is updated to $p^{(i)}(k+1)$ depending on the control action actually chosen and the reward (gain or loss) obtained.

<u>Step 4</u>: Depending on the control action chosen in step 2, the system in the meantime changes its state on $x(k+1)$. The cycle is now repeated until $p^{(i)}(k)$ converges to a unit vector.

In the present context $x(k)$ is <u>not known</u> at time k and this calls for a modification of the above procedure. Specifically in step 2 one does not know from which distribution the control action at time k is to be chosen. At each k only $z(n) = x(n-1)$ and $\hat{u}(n-1)$ are known. To get around this difficulty we propose to predict $x(n)$ based on the knowledge of $x(n-1)$, $u(n-1)$ and the transition probabilities corresponding to $x(n-1)$ and $u(n-1)$. It is here we use the explicit knowledge about P. If $\hat{x}(k) = x_j$ is the predicted value of $x(n)$, the control action $u(k)$ is chosen from the distribution $p^{(j)}(k)$. At time $(k+1)$, $z(k+1) = x(k)$, the true state of the system in the interval $(k, k+1]$ becomes known. Let $x(k) = x_m$. Now the modified procedure to be described below updates $p^{(m)}(k)$ to $p^{(m)}(k+1)$ depending on $x(k)$, $\hat{x}(k)$, $u(k)$ and the random reward.

In the sequel, we first describe a prediction algorithm for $\hat{x}(n)$. In general, prediction involves error and we will describe an algorithm that will minimize the probability of error.

<u>Prediction Algorithm</u>: If $z(k) = x(k-1) = x_i$, $\hat{u}(k-1) = \alpha_r$, then the state of the system at time k is controlled by the set of transition probabilities $p_{ij}[\alpha_r]$ $j = 1,2,\ldots,K$. Since P is known, these transition probatilities are known.

<u>Algorithm P1</u>: If $p_{i\ell}[\alpha_r] = \max_{j} \{p_{ij}[\alpha_r]\}$, then choose $\hat{x}(k) = x_\ell$. (Ties can be resolved at random).

For this algorithm P1,

$$\text{Probability of Error in Prediction} = \text{Prob}[x(k) \neq \hat{x}(k) \mid \alpha_r]$$

$$= \text{Prob}[x(k) \neq x_\ell \mid \alpha_r]$$

$$= \sum_{j \neq \ell} p_{ij}[\alpha_r]$$

Since $p_{i\ell}[\alpha_r]$ is the maximum of $p_{ij}[\alpha_r]$, for any other choice of $\hat{x}(k)$,
the probability of error will be more.

Learning Algorithm:

We shall now describe the basic steps in our learning algorithm.
Our description will focus on a typical interval $(k, k+1]$.

Step 1: At the beginning of the interval $(k,k+1]$, $x(k)$ is not known but
$z(k) = x(k-1)$ and $\hat{u}(k-1)$ are known exactly. These quantities are used in
the prediction algorithm P1. Let $\hat{x}(k) = x_i$.

Step 2: Since $\hat{x}(k) = x_i$, choose a control action from the distribution
$p^{(i)}(k)$. Let $\hat{u}(k) = \alpha_t$ be the control action chosen.

Step 3: At the end of the interval, that is, at time $k+1$, the actual
state $x(k+1) = x(k) = x_m$ (say) is made available.

The distributions are updated as follows:

$$p^{(m)}(k+1) = T^{i,m}[p^{(m)}(k), \hat{u}(k), H[x(k), \hat{u}(k)]]$$
$$p^{(j)}(k+1) = p^{(j)}(k) \quad (j \neq m) \tag{8.41}$$

where

$T^{i,m}: S_M \times U \times \{+1,-1\} \to S_M$ depends on both the actual and predicted states
at time k.

Step 5: Go to step 1 until all the $p^{(i)}(k)$, $i = 1$ to K converge to V_M.

In the following we describe a way in which the mapping $T^{i,m}$ can be
chosen:

Let $\hat{x}(k) = x_i$, $x(k) = x_m$, $\hat{u}(k) = \alpha_t$

    a)   If $x_i \neq x_m$, then $p^{(m)}(k+1) = p^{(m)}(k)$, that is, $T^{i,m}$ is essentially an identity mapping if $x_i \neq x_m$.

    b)   If $x_i = x_m$, then $T^{m,m}[., ., .]$ is described below:

$$p_t^{(m)}(k+1) = \begin{cases} p_t^{(m)}(k) + [1-p_t^{(m)}(k)]\,\lambda^{(m)}(k) & \text{if } H[k] = +1 \\ p_t^{(m)}(k) - [1-p_t^{(m)}(k)]\,\mu^{(m)}(k) & \text{if } H[k] = -1 \end{cases}$$

$$p_s^{(m)}(k+1) = \begin{cases} p_s^{(m)}(k) - p_s^{(m)}(k)\,\lambda^{(m)}(k) & \text{if } H[k] = +1 \\ p_s^{(m)}(k) + p_s^{(m)}(k)\,\mu^{(m)}(k) & \text{if } H[k] = -1 \end{cases} \qquad (8.42)$$
$$s \neq t$$

where $\lambda^{(m)}[.]$ and $\mu^{(m)}[.]$ satisfy the conditions (8C.1) - (8C.2). Notice the updating (8.42) follows the absolutely expedient algorithms described in Chapter 3. Further according to the above algorithm, there is updating only when there is correct prediction. Thus, there is basically an "inaction-update" feature built in the description of $T^{i,m}$.

    Analysis of this algorithm follows very closely the developments in Section 8.4 and we indicate the major results pertaining to this algorithm without proof. (Exercise 8.8)

    1) $\{h_m(k)\}$ is a submartingale for $m = 1,2,\ldots,K$. It follows from 8A'.2 that the system will visit any state from every state for any control action with a probability at least δ.

    2. $\lim_{k\to\infty} p^{(m)}(k)$ exists and belongs to V_M with probability one,

       $(m = 1,2,\ldots,K)$

    3) $p^{(m)}(k)$ changes only at those instances at which there is correct prediction. The subsequence $\{\tilde{p}^{(m)}(t)\}$ consisting of values of $\{p^{(m)}(k)\}$ at such instants constitute an imbedded stationary Markov process.

4) A lower bound on the probability with which $\tilde{p}^{(m)}(t)$ and hence $p^{(m)}(k)$ converges to each unit vector can be determined.

5) It can be shown by proper choice of the algorithm that the sequence of control vectors chosen corresponding to the algorithm (8.42) converges to the one step optimal control vector with probability $\geq 1 - \varepsilon$.

It is clear that our above algorithm crucially depends on the assumption that the matrices in the class P are known. When P is not known we need sophisticated prediction algorithm and this constitutes a major open problem on the topic of delayed state observation.

8.7 Comments and Historical Remarks

Control of a finite Markov chain whose transition probabilities and reward (or equivalently, cost) structure are known completely has a well established literature at the present time [B3][D1][K9][H8][V1]. The main thrust of the approach used in all these publications is to precompute the optimal control off line using the methods of dynamic programming so that they can be implemented on the system being controlled from the beginning.

More interesting is the control of a finite Markov chain with incomplete information. The ignorance may be contained in one or more of the

following: (1) unknown transition probabilities (2) unknown reward
structure and (3) noisy observation of the state. Astrom [A4] considered
the case of noisy observation of the state but assumed the knowledge of
the dynamics and reward structure. Silver and Moore[S8] calculate the
open loop optimal control when the state information is not available but
all the other relevant data are exactly known. An extensive treatment
of the control of a Markov chain with uncertain transition probabilities
is given by Martin [M5] who uses sequential updating of the transition
probability matrix in a Bayesian framework. Brooks and Leondes [B9] deal
with the problem of a fixed time lag in the state observation when all
the other quantities are known exactly. Eckles [E1] treats the optimal
maintenance problem with incomplete information. A common idea that
pervades most of these papers is to estimate the unknown quantity using
a Bayesian approach and then apply dynamic programming to obtain the
governing equations for the optimal controls. The cost structure is
always assumed to be known in all these publications. Such a procedure
becomes difficult to implement when the cost structure is unknown and in
these situations as illustrated in this chapter on-line learning techniques become very handy.

Tsetlin [T 2] was perhaps the first to consider the learning control
of a Markov chain. He applies the "play the winner" rule to a homogeneous finite state Markov chain whose transition probabilities are independent of the control action but cost depends both on the state and
control action. Both the transition matrix and the reward structure as a
function of the state and control action are unknown. Also it is assumed
in [T 2] that the state of the Markov chain is unobservable. That is,

the sequence of states visited by the Markov chain is <u>not known</u>. Further
details of the Tsetlin's paper are developed in exercise 8.9. Other related
papers in this context are Varshavskii and Vorontsova [V2], Lakshmivarahan
[L10']and Witten [W4]. Witten [W4] considers the cost to depend only on
state and not on control action.

An alternate viewpoint is to regard the problem as one of the adaptive
control models the controlled process by differential or difference
equation we use a Markov chain model. An advantage of such a viewpoint
is that certain noise effects form a natural part of the formulation and
further the assumption of a finite number of states yields a considerable
simplification. The results of this chapter can thus be regarded as an
application of learning theory to adaptive control. To be more specific,
it is an example of 'direct' adaptive control [N5] which bypasses the
need for explicit identification. Much of the development in this chapter
follows [L8].

It should be interesting to extend the results of this chapter to the
case of noisy state observations.

8.8 Exercises:

8.1) Under the assumption (8.7) prove (8.8) and (8.10).

8.2) For the stationary random Markov policy $R = \{\underline{u}\}$ show the following is true:

$$\lim_{k\to\infty} \Pi^T(k) = \lim_{k\to\infty} \Pi^T(0)\ P^k[u] = \Pi^T(\underline{u})$$

8.3) If $r = \{\underline{u}\}$, then using (8.10) derive (8.13) from (8.12).

8.4) Under the assumption 8A.2, show that $B_{ij}(k) \geq \delta$ for all $i,j = 1,2,\ldots,M$, $k \geq 1$ where $B(k) = [B_{ij}(k)]$ is defined in (8.21).

Hint: Define $B_j(k) = \min_i \{B_{ij}(k)\}$ and verify that

$$B_j(k+1) \geq B_j(k) \geq \ldots \geq B_j(0) \geq \delta > 0.$$

8.5) Derive the expression for $\delta h_i^i(k)$ as given in (8.29)

Hint: Same as exercise (3.10).

8.6) Show that $\tilde{p}^{(i)}(t)$ and $p^{(i)}(k)$ share all the asymptotic properties.

8.7) Prove theorem 8.3.

8.8) Prove all the properties of (8.42) listed at the end of section 8.6.

EPILOGUE

Having completed eight chapters it is now time to ask: "Are these the
only results on both the theory and applications that are known in the literature?"
Before answering this question we wish to remind the reader of the existence of
perpetual interaction (or feed back) between the theory and application in the
sense that an older theory giving rise to a new application which in turn gives
rise to newer theories and so on. In other words, it is extremely difficult in
general, to pin down where a theory ends and an application begins. With this
in mind, we wish to emphasize that "all most all" of the basic results on theory
of convergence of various classes of learning algorithms known at the time of
this writing are given in part I. On the application side however, our coverage
is by no means complete. The choice of topics for inclusion in part II was
essentially dictated by our familiarity. Yet for the benefit of the reader, in
the following, we refer to some of the other interesting applications.(a) Learning
approach to decentralised routing problem in telephone network has been extensively studied in Narendra, Wright and Mason [N6] Narendra and Thathachar
[N7] , Srikantakumar and Narendra [S14] , Srikantakumar [S15] and Kushner and
Huang [K14]. This constitutes one of the technological applications of learning
algorithms. For most of the classical results on network modelling and tele-
traffic routing refer Benes [B4] . (b) Learning approach to stochastic programming problems is given in Poznyak [P5] (c) Koditschek and Narendra [K15] and
Thathachar and Bhakthavatsalam [T9] deal with learning algorithms in multi-
teacher (or parallel) environments. (d) An analysis of the interaction between
a number of learning algorithms wherein the communication between different
algorithms is organized in a hierarchical or pyramidal structure is given in

Thathachar and Ramakrishnan [T8].

Except for [N6] and [N7], all the other papers mentioned in (a) are in the process of being published. Further most of the applications mentioned in (b) - (d) are in their formative stages. For a vast repertorie of potential applications of learning in general refer the classic work by Tsypkin [T4] [T5].

We hope this monograph will stimulate further research on both the theory and applications of learning algorithms.

REFERENCES

A1   Apostol, T.M., _Mathematical Analysis_. Addison Wesley, 1957.

A2   Arnold, L.   _Stochastic Differential Equations: Theory and Applications_ .
      Wiley, 1976.

A3   Aso, H. and Kimura M. "Absolute Expediency of Learning Automata". Infor-
      mation Sciences, Vol. 17, pp. 91-112, 1979.

A4   Astrom, K.J. "Optimal Control of Markov processes with incomplete state
      information". Journal of Mathematical analysis and applications. Vol. 10,
      pp. 174-205, 1965.

A5   Atkinson, R.C., Bower, G.H. and Crothers, E.J.   _An Introduction to_
      Mathematical Learning Theory   Wiley, 1965.

A6   Aumann, R.J. and Maschler, M.   "Repeated Games with incomplete infor-
      mation". Report ACD A/ST-116. Mathematica, Princeton, New Jersey, Vol. 4,
      pp. 7-24, 1968.

B1   Baba, N. and Sawaragi, Y., "On the learning behavior of stochastic automata
      under a non-stationary random environment. IEEE Transactions on systems,
      man and cybernetics, Vol. 5, pp. 273-275, 1975.

B2   Bechhofer, R.E., Kiefer, J. and Sobel, M.   _Sequential identification and_
      ranking procedures . The University of Chicago Press, Chicago, 1968.

B3   Bellman, R.E. _Adaptive Control Processes_. _A Guided Tour_, Princeton
      University Press, 1961.

B4   Benes, V., _Mathematical Theory of Connecting Networks and Telephone_
      Traffic. Academic Press, 1965.

B5   Billingsly, P. _Convergence of Probability Measures_, Wiley, 1968.

B6   Blaquiere, A. _Nonlinear System Analysis_. Academic Press (Chapter 3). 1966.

B7   Blum, J. "Approximation methods which converge with probability one. Annals of Mathematical Statistics. Vol. 25, pp. 382-386, 1954.

B8   Bradt, R.N., Johnson, S.M. and Karlin, S. "On sequential designs for maximizing the sum of n-observations". Annals of Mathematical Statistics. Vol. 27, pp. 1060-74, 1956.

B9   Brooks, D. and Leondes, C. "Markov Decision processes with state information lag". Operations Research, Vol. 20, pp. 904-907, 1972.

B10   Bush, R.R. and Mosteller, F. "Stochastic Models for Learning" Wiley 1958.

B11   Brown, G.W. "Iterative solutions of games by fictitious play". in Activity Analysis of Production and Allocation (Ed) T.C. Koopmans, Cowles Commission Monograph 13, 1951, pp. 374-376.

C1   Chandrasekaran, B. and Shen, D.W.C. "On expediency and convergence in Variable Structure Automata". IEEE Transaction Systems Science and Cybernetics. Vol. 4, pp. 52-60, 1968.

C2   Chandrasekaran, B. "Stochastic Automata Games" IEEE Transactions on Systems Science and Cybernetics. Vol. 5, pp. 145-146, 1969.

C3   Chester, C.R. Techniques in Partial Differential Equations McGraw Hill, 1971. Chapter 1 and 8.

C4   Cockrell, L.D. "On Search Techniques in Adaptive Systems." Dissertation, Purdue University, Lafayette, Indiana, 1970.

C5   Courant, R. and Hilbert, D. Methods of Mathematical Physics-Vol. II, 1962. Chapter 1 and 2.

C6   Cover, T.M. and Hellman, M.E., "Two Armed Bandit Problem with Time-Invariant Finite Memory", IEEE Transactions on Information Theory, Vol. 14, pp. 185-195 1970.

C7   Cover, T.M. "A Note on the Two Armed-Bandit Problem with Finite Memory".
      Information and Control, Vol. 12, pp. 371, 1968.

C8   Cox, D.R. and Miller, H.D. The Theory of Stochastic Processes, Wiley,
      1965.

C9   Crawford, V.P. "Learning the Optimal Strategy in a Zero-Sum Game."
      Econometrica, Vol. 42, pp. 885-891, 1974.

C10   CSIBI.S Stochastic Processes with Learning Properties . Springer Verlag
      1975.

D1   Derman, C. "Markovian Decision Processes. Average Cost Criterion."
      in Mathematics of Decision Sciences (Ed) by Dantzig and Vienott. A.M.S.
      Publications, pp. 139-148, 1968.

D2   Doob, J.L.   Stochastic Processes   Wiley, 1951.

D3   Dunford, N. and Schwartz, J.T. Linear Operators, Vol, I, Wiley, 1966.

E1   Eckles, J.E. "Optimum Maintenance with Incomplete Information". Operations
      Research, Vol. 16, pp. 1058-1067, 1968.

E2   El-Fattah, Y.M. "Stochastic Automata Modelling of Certain Problems of
      Collective Behavior". IEEE Transactions on Systems, Man and Cybernetics.
      Vol. 10, pp. 304-314.

F1   Feldman, D. "Contributions to the Two-Armed Bandit Problem". Annals, of
      Mathematical Statistics, Vol. 33, pp. 817-856, 1962.

F2   Flerov, Yu. A., Some Classes of Multi Input Automata". Journal of Cybernetics.
      Vol. 2, pp. 112-122, 1972.

F3   Fox, B.L., "Finite Horizon Behavior of Policies for Two-Armed Bandits".
      Journal of American Statistical Association. Vol. 69, pp. 963-965, 1974.

F4   Fu, K.S. "Learning Control Systems - Review and Outlook". IEEE
      Transactions on Automatic Control, Vol. 15, pp. 210-221, April, 1970.

F5   Fu, K.S., Pattern Recognition and Machine Learning , Plenum Press, 1971.

F6   Fu, K.S., "Learning Control Systems and Intelligent Control Systems: An
      Intersection of Artificial Intelligence and Automatic Control." IEEE
      Transactions on Automatic Control, Vol. 16, pp. 70-72, 1971.

F7   Fu, K.S., "Stochastic Automata Models for Learning Systems". in Computers
      and Information Sciences II (Ed) by J.T. Tou, Academic, 1967.

F8   Fukunaga, K. Introduction to Statistical Pattern Recognition, Academic
      Press, 1972.

G1   Gel'fand, I.M., Pyatetskii-Shapiro I.I. and Tsetlin, M.L. "On Certain
      Classes of Games and Games of Automata". Doklady Akademy Nauka, Vol. 152,
      1963.

G2   Gladyshev, E.G. "On Stochastic Approximation." Theory of Probability
      and its Applications. Vol. 10, pp. 275-278, 1965.

G3   Gikham, I.I. and Skorokhod, A.V. Introduction to Theory of Random
      Processes . W.B. Saunders Company, 1965.

G4   Gikhman, I.I. and Skorokhod, A.N., Stochastic Differential Equations .
      Springer-Verlag, 1972.

G5   Ginsburg, S.L., Krylov, V. Yu and Tsetlin, M.L. "On One Example of a
      Game of Many Identical Automata". Automation and Remote Control, Vol, 25,
      pp. 608-612, 1964.

H1   Harsanyi, J.C., "Games with Incomplete Information Played Bayesian Players -
      I." Management Science, Vol. 14, pp. 159-182, 1967.

H2   Herkenrath, U. and Theodorescu, R. "General Control Systems". Information
      Sciences, Vol. 14, pp. 57-73, 1978.

H3   Herkenrath, U., Kalin, D., Lakshmivarahan, S. "A General Class of
       Absorbing Barrier Learning Algorithms". Information Sciences, 1981
       (To Appear) Also EECS Technical Report, University of Oklahoma, March,
       1980.

H4   Hilgard, E.R., Theories of Learning Appleton-Century-Crofts, Inc., 1948.

H5   Ho, Y.C. and Chu, K.C. "Information Structure in Dynamic Multiperson
       Control Problem". Automatica, Vol. 10, pp. 341-352, 1974.

H6   Ho, Y.C., Kastner, M.P. and Wong, E. "Team Signalling and Information
       Theory." IEEE Transactions on Automatic Control, Vol. 23, pp. 305-312, 1978.

H7   Hoel, D.A., Sobel, M., Weiss, G.H., "A Two Stage Procedure for Choosing
       the Better of Two Binomial Populations". Biometrica, Vol. 59, pp. 317-322,
       1972.

H8   Howard, R.A., Dynamic Programming and Markov Processes . M.I.T. Press,
       1960.

H9   Hull, C.L. Principles of Behavior, Appleton-Century-Crofts, Inc., 1943.

I1   Iglehart, D.L. "Weak Convergence in Applied Probability". Stochastic
       Processes and Their Applications. Vol. 2, pp. 211-241, 1974.

I2   Iosifescu, M. and Theodorescu, R.,   Random Processes and Learning
       Springer Verlag, 1969.

I3   Isbell, J.R., "On a Problem of Robbins". Annals of Mathematical Statistics,
       Vol. 30, pp. 606-610, 1959.

J1   Jarvis, R.A. "Adaptive Global Search in a Time Varying Environment Using
       Probabilistic Automaton with Pattern Recognition Supervision". IEEE
       Transactions Systems, Science and Cybernetics. Vol. 6, pp. 209-217, 1970.

K1 Kemeny, J.G. and Snell, J.L. <u>Finite Markov Chains</u> Von Nostrand, 1960.

K2 Kohlberg, E. "Optimal Strategies in Repeated Games With Incomplete Information:, International Journal of Game Theory, Vol. 2, pp. 99-110, 1974.

K3 Krinskii, V.I., "An Asymptotically Optimal Automaton With Exponential Convergence". Bio Physics, Vol. 9, pp. 484-487, 1964.

K4 Krylov, V. Yu. "On One Stochastic Automation Which is Asymptotically Optimal in a Random Media". Automation and Remote Control, Vol, 24, pp. 1114-1116, 1963.

K5 Krylov, V. Yu and Tsetlin, M.L. "Games Between Automata". Automation and Remote Control, Vol. 24, pp. 889-900, 1963.

K6 Kurtz, T.G. "Solutions of Ordinary Differential Equations as Limits of Pure Jump Markov Process". Journal of Applied Probability, Vol. 7, pp. 49-58, 1970.

K7 Kurtz, T.G. "Limit Theorems For Sequences of Jump Processes Approximating Ordinary Differential Processes". Journal of Applied Probability, Vol. 8, pp. 344-356, 1971.

K8 Kushner, H.J., Thathachar, M.A.L. and Lakshmivarahan, S. "Two-State Automation With Linear Reward-Inaction Reinforcement Scheme - A Counter Example". Department of E.E. Indian Institute of Science. Bangalore, India. TR-EE 1970. (Also appeared in IEEE Transactions on Systems, Man and Cybernetics Vol. 2, pp. 292-294, 1972).

K9 Kushner, H.J. <u>Introduction to Stochastic Control</u> Holt, Rinehart and Winston, New York, 1971.

K10   Kushner, H.J. and Clark, D.S.   Stochastic Approximation for Constrained
       and Unconstrained Systems   Springer Verlag, 1978

K11   Kushner, H.J. "General Convergence Results for Stochastic Approximations
       Via Weak Convergence Theory". LCDS Technical Report 76, Brown University,
       Providence, Rhode Island, August, 1976.

K12   Kushner, H.J. "Convergence of Recursive Adaptive and Identification
       Procedures Via Weak Convergence Theory". IEEE Transactions on Automatic
       Control. Vol. 22, pp. 921-930, 1977.

K13   Kushner, H.J.   Probability Methods for Approximations in Stochastic
       Control and for Elliptic Equations. Academic Press, 1977.

K14   Kushner, H.J. and Huang Hai, "Averaging Methods for the Asymptotic
       Analysis of Learning and Adaptive Systems with Small Adjustment Rate".
       LCDS Technical Report 80-1. Brown University. April, 1980.

K15   Koditschek, D.E. and Narendra, K.S. "Fixed Structure Automata in Multi-
       teacher Environment". IEEE Transactions on Systems, Man and Cybernetics,
       Vol. 7, pp. 616-624, 1977.

L1   Lakshmivarahan, S. and Thathachar, M.A.L. "Optimal Non-Linear Reinforce-
       ment Schemes for Stochastic Automata". Information Sciences, Vol. 4,
       pp. 121-128, 1972.

L2   Lakshmivarahan, S. and Thathachar, M.A.L. "Absolutely Expedient Learning
       Algorithms for Stochastic Automata". IEEE Transactions Systems Man and
       Cybernetics, Vol. 3, pp. 281-286, 1973.

L3   Lakshmivarahan, S. and Thathachar , M.A.L. "Absolute Expediencey of Q- and
       S- Model Learning Algorithms. IEEE Transactions Systems Man and Cybernet-
       ics. Vol. 6, pp. 222-226, 1976.

L4   Lakshmivarahan, S. "Learning Algorithms for Stochastic Automata".
Ph.D. Thesis. Indian Institute of Science, Bangalore, India, 1973.

L5   Lakshmivarahan, S. and Thathachar, M.A.L. "Bounds On The Probability of
Convergence of Learning Automata". IEEE Transactions Systems Man and
Cybernetics, Vol. 6, pp. 756-753, 1976.

L6   Lakshmivarahan, S. and Narendra, K.S. "Learning Algorithms for Two
Person Zero Sum Stochastic Games With Incomplete Information". Mathematics of Operations Research. Vol. 6, No. 4, 1981.

L7   Lakshmivarahan, S. "ε-Optimal Learning Algorithms- Non-absorbing Barrier
Type". Technical Report EECS 7901. Feb. 1979, School of Electrical
Engineering and Computing Sciences, University of Oklahoma, Norman,
Oklahoma.

L8   Lakshmivarahan, S. and Thathachar, M.A.L. "On-Line Learning Control
of a Markov Chain with Unknown Dynamics and Cost-Structure" Part I
and Part II., School of Electrical Engineering and Computing Sciences,
University of Oklahoma. Tech. Report 7902, July, 1979.

L9   Lakshmivarahan, S. and Narendra, K.S. "Learning Algorithms for Two-Person
Zero Sum Stochastic Games With Incomplete Information -- A Unified
Approach". Becton Center Technical Report, Yale University S & I.S. TR

L10   Lakshmivarahan, S. "Two Person Decentralized Team with Incomplete
Information". Applied Mathematics and Computation, Vol. 8, pp. 51-78,
1981.

L10'   Lakshmivarahan, S. "Learning Algorithms for Stochastic Automata Acting
In Non-stationary Random Environments". Journal of Cybernetics. Vol. 4,
pp. 73-85, 1974.

L11   LaSalle, J.P. and Lefshetz, S.   Stability by Liapunov's Direct Method.
Academic Press, New York, 1961.

L12   Ljung, L. "Analysis of Recursive Stochastic Algorithms". IEEE Transactions
On Automatic Control, Vol. 22, pp. 551-575, 1977.

L13   Loeve, M. Probability Theory Von Nostrand, 1963.

L14   Luce, R.D. Individual Choice Behavior , Wiley, 1959.

L15   Luce, R.D. and Raiffa, H. Games and Decisions . Wiley, 1957.

M1   Martin, J.J. Bayesian Decision Problems and Markov Chains , Wiley, 1967.

M2   Marshak, J. and Radner, Roy. Economic Theory of Teams, Yale University
Press, 1972.

M3   McLaren, R.W. "A Stochastic Automaton Model for a Class of Learning
Controllers". Joint Automatic Control Conference, Preprints, 1967.

M4   McMurtry, G.J. and Fu, K.S. "A Variable Structure Automaton Used as a
Multi-Model Search Technique" IEEE Transactions on Automatic Control,
Vol. 11, pp. 379-387, 1966.

M5   Meerkov, S.M. "Simplified Description of Slow Markov Walks" Part I.
Automatika i Telemechanika No. 3, pp. 66-75, 1972.

M6   Mendel, J.M. "Reinforcement Learning Models and Their Applications
to Control Problems: Learning Systems". 1973 Joint Automatic Control
Conference Proceedings.

M7   Mendel, J.M. and Fu, K.S. Adaptive, Learning and Pattern Recognition
Systems . Academic Press, 1970.

N1   Narendra, K.S. and Thathachar, M.A.L. "Learning Automata - A Survey"
IEEE Transactions Systems Man and Cybernetics, Vol. 4, pp. 323-334, 1974.

N2   Narendra, K.S. "Learning Models Using Stochastic Automata". Proceeding
       of 1972 International Conference on Cybernetics and Society. Washington,
       D.C. 1972.

N3   Narendra, K.S. and Thathachar, M.A.L.   Learning Automata   (forthcoming
       book).

N4   Narendra, K.S. and Lakshmivarahan, S. "Learning Automata - A Critique"
       Journal of Cybernetics and Information Sciences - Special Issue on Learning
       Automata. Vol. 1, pp. 53-66, 1978.

N5   Narendra, K.S. and Valavani, L.S. "Direct and Indirect Adaptive Control".
       Becton Center S & I S Report-7711, Yale University, New Haven, Connecticut

N6   Narendra, K.S., Wright, E. and Mason, L.G. "Application of Learning
       Automata to Telephone Traffic Routing." IEEE Transactions on Systems
       Man and Cybernetics, pp. 785-792, 1977.

N7   Narendra, K.S. and Thathachard, M.A.L. "On the Behavior of a Learning
       Automaton in a Changing Environment with Routing Applications." IEEE
       Transactions on Systems Man and Cybernetics, Vol. 10, pp. 262-269, 1980.

N8   Nevelson, M.B. and Has'minskii, R.Z.   Stochastic Approximation and
       Recursive Estimation   Translation of the American Mathematical Society
       Vol. 47, 1973.

N9   Nilsson, N.J.   Problem Solving Methods in Artificial Intelligence .
       McGraw Hill, 1971.

N10   Norman, M.F. Markov Processes and Learning Models, Academic Press, 1972.

N11   Norman, M.F. "On Linear Models With Two Absorbing Barriers", Journal of
       Mathematical Psychology. Vol. 5, pp. 225-241, 1968.

N12   Norman, M.F. "Some Convergence Theorems for Stochastic Learning Models
       with Distance Diminishing Operators". Journal of Mathematical Psychology.
       Vol. 5, pp. 61-101, 1968.

N13   Norman, M.F. "A Central Limit Theorem for Markov Processes that Move by Small Steps". The Annals of Probability. Vol. 2, pp. 1065-1074, 1974.

N14   Norman, M.F. "Slow Learning". The British Journal of Mathematical and Statistical Psychology. Vol. 21, pp. 141-159, 1968.

N15   Norman, M.F. "Markovian Learning Process" SIAM Review, Vol. 16, pp. 143-162, 1974.

N16   Norman, M.F., "Approximation of Stochastic Processes by Gaussian Diffusions, and Applications to Wright-Fisher Genetic Model." SIAM, Journal of Applied Mathematics, Vol. 29, pp. 225-242, 1975.

P1   Parthasarathy, K.R. <u>Probability Measures in Metric Spaces</u>, Academic Press, 1965.

P2   Ponomarev, V.A. "A Construction of an Automaton which is Asymptotically Optimal in a Stationary Random Media". Bio Physics, Vol. 9, pp. 104-110, 1964.

P3   Ponssard, J.P. and Zamir, S. "Zero Sum Sequential Games With Incomplete Information". International Journal of Game Theory, Vol. 2, pp. 99-110, 1974.

P4   Poznyak, A.S. "Investigation of Convergence of Algorithms for Learning Stochastic Automata". Automation and Remote Control, pp. 77-91, 1973.

P5   Poznyak, A.S. "Learning Automata in Stochastic Programming Problem". Automation and Remote Control. pp. 1608-1619, 1973.

R1   Robbins, H. "Sequential Decision Problem With Finite Memory". Proceedings of the National Academy of Sciences, Vol. 42, pp. 920-923, 1956.

R2   Robbins, H. "Some Aspects of the Sequential Design of Experiments". Bulletin of the American Mathematical Society. Vol. 58, pp. 529-532, 1952.

R3   Robbins, H. and Monro, S. "A Stochastic Approximation Method." Annals or Mathematical Statistics. Vol. 22, pp. 400-407, 1951.

R4   Robbins, H. and Siegmund, D.O. "Sequential Tests Involving Two Populations"
      Journal of American Statistical Association. Vol. 69, pp. 132-139, 1974.

R5   Rosen, B. "On the Central Limit Theorems for Sums of Dependent Random
      Variables". Z. Wahrsch. und Verw. Gebiete, Vol. 7, pp. 48-82, 1967.

R6   Robinson, J. "An Iterative Method of Solving a Game". Annals of Mathe-
      matics, Vol. 54, 1951, pp. 296-301.

S1   Samuels, S.M. "Randomized Rules for the Two-Armed Bandit with Finite
      Memory". Annals of Mathematical Statistics. Vol. 39, pp. 2103-2107,
      1968.

S2   Sanghvi, A.P. and Sobel, M.J. "Bayesian Games as Stochastic Processes".
      International Journal of Game Theory. Vol. 5, pp. 1-22, 1976.

S3   Saridis, G.N. "On-Line Learning Control Algorithms - Learning Systems",
      1973 Joint Automatic Control Conference.

S4   Saridis, G.N.   Self-Organizing Control of Stochastic Systems . Marcel-
      Dekker Inc. 1978.

S5   Sawaragi, Y. and Baba N. "Two ε-Optimal Non-linear Reinforcement Schemes
      for Stochastic Automata." IEEE Transactions on Systems, Man and Cyber-
      netics, Vol. 4, pp. 126-131, 1974.

S6   Scarf, H.E. and Shapley, L.S. "Games with Partial Information" in
      Contributions to the Theory of Games. Vol III, (Ed) Kuhn and Tucker,
      Princeton University Press. 1957, pp. 213-231.

S7   Shapiro, I.J. and Narendra, K.S. "Use of Stochastic Automata for Parameter
      Self-Optimization with Multimodal Performance Criteria". IEEE Transactions
      on Systems Man and Cybernetics. Vol. 5, pp. 352-360, 1969.

S8   Silver, E.A. and Moore, J.B. "Mixing of Markov Processes". Decision
      Sciences, Vol. 7, pp. 383-93, 1976.

S9   Sklansky, J. "Learning Systems for Automatic Control", IEEE Transactions
      on Automatic Control, Vol. 11, pp. 6-19, 1966.

S10   Slagle, J.   Artificial Intelligence and Heuristic Programming .
      McGraw Hill, 1975.

S11   Smith, C.V. and Pyke, R. "The Robbins-Isbell Two Armed Bandit Problem
      With Finite Memory". Annals of Mathematical Statistics. Vol. 36,
      pp. 1375-1386, 1965.

S12   Sobel, M. and Weiss, G.H. "Play the Winner Rule and the Inverse Sampling
      for Selecting the Best of K $\geq$ 3 Bionomial Populations". Annals of
      Mathematical Statistics. Vol. 43, pp. 1808-1826, 1972.

S13   Solomonoff, R. "Some Recent Work in Artificial Intelligence". Proceedings
      of IEEE, Vol. 54, Dec. 1966.

S14   Srikantakumar, P.R. and Narendra, K.S. "Learning Algorithm Model for
      Routing in Telephone Networks". Becton Center S & I.S. Report 7903.
      Yale University, May, 1979.

S15   Srikantakumar, P.R., "Learning Models and Adaptive Routing in Communications
      Network." Ph.D. Dissertation Yale University, August, 1980.

S16   Sternberg, S. "Stochastic Learning Theory", in Handbook of Mathematical
      Psychology. (Ed) Luce, Bush and Galanter, Vol. II. John Wiley, pp. 1-120,
      1963.

S17   Suppes, P. and Atkinson, R.C.   Markov Learning Models For Multiperson
      Interaction . Stanford University Press, 1960.

T1   Thompson, W.R. "On the Likelihood That One Unknown Probability Exceeds
      Another in View of the Evidence of Two Samples". Biometrica, Vol. 25,
      pp. 285-294, 1933.

T2   Tsetlin, M.L., Automaton Theory and Modelling of Biological Systems .
      Academic, 1973.

T3   Tsetlin, M.L., "On the Behavior of Finite Automata in Random Media".
      Automation and Remote Control. Vol. 22, pp. 1345-1354, 1961.

T4   Tsypkin, Ya. Z. Adaptation and Learning in Automatic Systems . Academic
      1971.

T5   Tsypkin, Ya. Z. Foundations of the Theory of Learning Systems . Academic
      Press, 1973.

T6   Tsypkin, Ya. Z. and Poznyak, A.S. "Finite Learning Automata", Engineering
      Cybernetics, Vol. 10, pp. 478-490, 1972.

T7   Tucker, A.W. A Graduate Course in Probability Theory . Academic Press,
      1972.

T8   Thatachar, M.A.L. and Ramakrishnan, K.R. "A Hierarchical System of
      Learning Automata". EE Report 57, Jan. 1980, Indian Institute of Science,
      Bangalore, India.

T9   Thatachar, M.A.L. and Bhakthavathsalam, R. "Learning Automaton Acting
      in Parallel Environments". Journal of Cybernetics and Information Sciences.
      Vol. I, pp. 121-127, 1977.

V1   Variya, P. "Optimal and Sub Optional Stationary Controls for Markov
      Chains". IEEE Transactions on Automatic Control Vol-AC 23, pp. 338-394,
      1978.

V2   Varshavskii, V.I. and Vorontsova, I.P. "On the Behavior of Stochastic
      Automata with Variable STructure". Automation and Remote Control,
      Vol. 24, pp. 327-333, 1963.

V3   Viswanathan, R. and Narendra, K.S., "Expedient and Optimal Variable
      Structure Stochastic Automata". Dunham Lab. TR-37, 1970. Yale University,
      New Haven, Connecticut.

V4   Viswanathan, R. and Narendra, K.S. "Games of Stochastic Automata".
IEEE Transactions Systems Man and Cybernetics, Vol. 4, pp. 131-135, 1974.

V5   Viswanathan, R. and Narendra, K.S. "Stochastic Automata Models with
Application to Learning Systems". IEEE Transactions Systems Man and
Cybernetics, Vol. 3, pp. 107-111, 1973.

V6   Viswanathan, R. and Narendra, K.S., "A Note on Linear Reinforcement
Scheme for Variable Structure Stochastic Automata". IEEE Transactions
on Systems, Man and Cybernetics. Vol. 2, pp. 292-294, 1972.

V7   Von Neumann, J. and Morgenstern, O. Theory of Games and Economic
Behavior , Princeton University Press, 1947.

V8   Vorentsova, I.P. "Algorithms for Changing Automaton Transition
Probabilities". Problemy Peredachi Informatsii, Vol. 1, pp. 122-126,
1965.

W1   Wasan, M.T. Stochastic Approximations. Cambridge University Press, 1969.

W2   Witten, I.H. "Finite Time Performance of Some Two-Armed Bandit Controller".
IEEE Systems Man and Cybernetics. Vol. 3, pp. 194-197, 1973.

W3   Witten, I.H. "The Apparent Conflict Between Estimation and Control - A
Survey of Two-Armed Bandit Problem". Journal of Franklin Institute,
Vol. 301, pp. 161-189, 1976.

W4   Witten, I.H. "An Adaptive Optimal Controller for Discrete-Time Markkov
Environments". Information and Control. Vol. 34, pp. 286-295, 1977.

Z1   Zamir , S. "On the Notion of the Value for Games with Infinitely Many
Stages". Annals of Statistics, Vol. 1, pp. 791-796, 1973.

INDEX